G. Roger Stair

Sasquatch and Me

17+ Years of Private Research Revealed

G. Roger Stair

"There's a way to do it better—find it."

—Thomas Edison

G. Roger Stair

ISBN-13: 978-1-7321712-2-0

"You only need sit still long enough in some attractive spot in the woods that all its inhabitants may exhibit themselves to you by turns."

— Henry David Thoreau, 12. Brute Neighbors, *Walden*

Dedication

To the love of my life, my partner, my dear wife Maggie Ann, our six adult children and their spouses, and, of course, our nine delightful, sometimes brilliant, sometimes sassy grandchildren. Love to each and all of you! I hope this book does not embarrass you too much!

Table of Contents

Disclaimer

This book is presented solely for educational and entertainment purposes. *Sasquatch and Me* is not a scientific tome; rather, it is the story of how I inadvertently stepped into the realm of private research for the so-called sasquatch. While best efforts have been used in preparing this book, the author and publisher make no representations or warranties of any kind and assume no liabilities of any kind with respect to the accuracy or completeness of the contents and specifically disclaim any implied warranties or fitness of use of the material herein for a particular purpose.

If errors exist, they are not intentional. If significant errors are found after print, a second edition will correct them. Neither the author nor the publisher shall be held liable or responsible to any person or entity with respect to any loss or incidental or consequential damages caused, or alleged to have been caused, directly or indirectly, by the information, suggestions, or techniques contained herein.

I may not have implemented all suggestions made by the editor. For example, there may be times I chose to keep an original passage or text for context or emphasis. Or, I may have used a form or format that appealed to me and not the editor. Regardless, this relieves the editor from responsibility if the suggestions made were not implemented.

Acknowledgments

I offer a heartfelt "thank you" to all who so graciously, generously contributed over the years to the unfoldment of research presented within, and to those who played a direct role in relentlessly reviewing, editing, and publishing the final version of this book.

There are far too many people who contributed to either the research or directly to this book to name them all. Even so, a number **must** be named and duly recognized.

To Marilyn and Ken for years of gracious hospitality to so many. You opened your doors to strangers, many who became close friends. Our love runs deep for you, Marilyn, and we are forever indebted to you. Remember the year I stopped myriad times across the Upper Peninsula, so you and Maggie could garage sale and thrift shop? What time did we get home? Two or three a.m.? Yikes! We sure were tired, eh? How about the time I took you and Maggie out at night and the two of you danced in the *faerie* lights? Or ....

To Joanie, who stood staunchly alone when others ran; words are totally inadequate to even begin to tell you how much you mean to Maggie and me. You are forever in our hearts and we send heartfelt love to you daily. I brought this project to fruition, in part, due to your loving, and sometimes annoying, encouragement. Without your love and support, none of this would have happened! May the angels who grant blessings surround you always, and always grant you the blessings you so richly deserve—which are way too many to count.

To Steve, Troy, and, of course, Sue. What to say? You were there at the start; too, you were there for years following. Oh, the hikes in the wilderness, the campsites (good and bad), the hours and hours of thought-provoking discussions while sitting around campfires listening to nonsensical sounds emanating from the wilderness. Each of you deserve huge hugs and enormous thanks for all you've contributed to the whole of this research. I never really knew where this would end. I certainly couldn't have foretold a book; yet, here it is. I love you and thank you from the depths of my heart.

To Kirk and Karen—the woman who brought you happiness and joy. I remember one year (a lifetime ago) it was Kirk, Jack, Maggie, and me. Goodness! The hours of driving, hiking, wandering day and night added up to amazing experiences, long talks, and much speculation. So many strange sights and sounds. Thanks so much for being our friends. I remember the Fall before I moved the research to northern Wisconsin when Kirk, Marilyn, Maggie, and I drove to an area I had found on one of my many maps to check for signs. We parked alongside the road, got out, and stepped into the woods. We at once saw prints and more prints and more prints. You looked at me and said, "Damn!" Well, Kirk and Karen, "Damn! The two of you are awesome and true blue!"

To Al and Pat for listening to my story of missing dirt, and especially to Al for your generous offer to have my dirt tested. The results of that test allowed me to approach this mysterious quandary with renewed vigor along with a clearer understanding for the next appropriate step in this ongoing saga of sasquatch and me. You are greatly appreciated!

To Dawn and Shawn for use of the truck and listening to my never-ending nonsensical babblings about my research. My goodness! We sure are blessed with generous-hearted children. My research would have been so different and far less without you letting me borrow the truck over the years. I love you more than I can say. Thank you so very much! Oh, by the way, I haven't forgotten the story you told me about Bob and Tom, those radio goofs who seem to have tickled you so much you laughed yourselves to tears on your way to work early one morning. Seriously! What in the world could be so funny about Bob and Tom asking on air, "What would you do if you found out your dad was searching for sasquatch?" Hope you youngsters were late for work that morning! Smile.

To April, another of our amazing children, and her husband Steve. April, you loaned me your truck to haul my trailer north. I love that Maggie and I have daughters who love trucks! Hmm! Would you have let me take it knowing what you know now? I'm guessing yes, because you're the bestest! You have embedded yourself within my heart, April, and I love you very much! You are definitely worthy of praise!

To Beth, and her husband Stephen, for use of the kayaks over the years. You had a hint, Stephen, about what I was doing and still let me take your kayaks! Silly boy! A sasquatch could have whittled your kayaks into plastic toothpicks! Beth, how many times did you drive Mitchell or Sophia to our house to mow, vacuum and take care of our cats? Too many to count! Beth and Stephen, your gracious generosity will never be forgotten, and you are deeply loved.

To Andrew, who went to the wilderness with grandma and me. You drove grandma up and back, allowing me to continue my research after you left. I love you, Andrew, and am proud of who you have become! Sherry, Andrew's mom and another daughter, joined us during an early heart-head research excursion. Oh, my! The stories she can tell! Kristin, another daughter, along with Brandon and Ashleigh, a couple more grandchildren, have yet to experience our wilderness haven.

To two more of our grandchildren, Matt and Amber, for the many hours you each spent looking after the house, cats, and lawn while I was wandering through the wilderness searching for a silly legend! And to Amber and Sophia for helping to maintain my home away from home, my tried and true trailer. I love you each more than I can say and thank you from the bottom of my heart.

To Chad and Jennifer for letting Zack and Nick help keep the lawn in shape while grandpa chased a myth through the wilds of the Chequamegon-Nicolet National Forest. Every bit of help that came my way meant that I was able to keep pursuing my desire of meeting and interacting with sasquatches. Go figure! The truth is now out!

To Val, who agreed to the grueling task of editing this tome. I am deeply honored you agreed to this, though I might question the sanity of your decision, and I surely send you my love daily! Thanks so much for being my friend and for sharing a part of your life with me. Also, thank you for your patient and loving acceptance of my stories about my research. You have a special place in my heart.

To Jules, who sacrificed so very much to aid me in my quest of understanding sasquatches; sore arms, sore shoulders, sore back, sore knees, sorely tired at times and the vacation days (weeks) used over the years, which included a sore wallet! Through the ups and downs, the doubts and fears, the immense joy and encouragement we felt accompanying success, you stood always trustworthy and supportive. Those times I doubted myself, you offered encouragement. When I decided I was nuts and thought I should quit, you said something like, "What's wrong with nuts?" Whether I was strong or weak in the research, you stood steadfastly beside me. In other words, you stayed the course, all the way to where we are now. Stated simply, thank you so much and know that I love you. And please do remember, we are far from done! Go yoga! Oh, and do not get rid of the "Woody" muck boots.

Finally, to Maggie. I sorely miss the days when you walked beside me through the wilderness. Without your years of love and ongoing support, I would not be who I am today; nor would this book have ever reached the print stage. Every person we meet throughout life plays a role in shaping and molding the whole of who we are as conscious humans. A select few play roles no other could. Aside from the myriad hours of reading and rereading this manuscript, the overall role you've played in my life since July 20, 1990 fits neatly into the category of **no other could**. We've been through it together, and together we remain; said simply, I love you.

So many more aided the research process over the years. If you ever read this, simply think back and you will know that I send each of the unnamed a heartfelt thanks.

—grs

G. Roger Stair

Foreword

I started off as a high school biology and chemistry teacher. During that time, I took a couple of groups of students to camp in the backcountry of Quetico in Ontario, Canada. The Ontario Parks website describes Quetico as, "… an iconic wilderness class park renowned for its rugged beauty, towering rock cliffs, majestic waterfalls, virgin pine and spruce forests, picturesque rivers and lakes."[1] Those backcountry trips deepened my appreciation of nature. I've always enjoyed being outside, however, being in the wilderness is inexplicable.

I first met Roger when I took classes that he taught at a local institute in 1997. Little did I know of the friendship that would ensue and the journey we'd experience as the years unfolded. He has many notable and amazing qualities that I admire. One of which is his passion for exhaustively researching prior to presenting on a topic. *No source* escapes investigation, and his unwavering integrity, coupled with his love of research, are core traits he embodies that play out in this book. As you will see, he also has a unique sense of humor.

I took part in some of his early heart-head research trips and absolutely loved every facet! Many of the experiences we had during the early years tapped into and awakened a part of me so relevant to the core of my being. Roger was patient to say the least. We were all as curious as kids wanting to experience and understand the impact that Earth's changing magnetics had on us. The wilderness was gorgeous and rejuvenating.

Fortunately, our personalities, mutual respect and appreciation for each other—as well as our deep respect for the wilderness—meshes, and we work well together in the wilderness and around camp. We always discuss plans for each day, and often must adjust due to weather changes or unexpected snafus of equipment that would keep us at camp. The days we are confined to camp also contribute to the research, as we often find evidence that we had visitors overnight. As enjoyable and enlightening as documenting that evidence is, the conversations and connection we experience while sharing thoughts, ideas and stories is both research probative and instructive.

Roger expertly, innately and impressively knows how to safely navigate walking, kayaking, or canoeing through primitive country. His inherent interest and extensive background in electronics, magnetism and his research into its effects upon humans and nature led him to his first encounter with a sasquatch in the western Upper Peninsula of the state of Michigan. His eclectic background, as well as his diligence and commitment to the practice of achieving and sustaining a state of coherence, combine with his knowing of the importance of coherence; this superbly aligns as he tells his story of nearly two decades of private sasquatch research. Roger is his truest self when he is in his element researching—passionate, focused, hardworking, and silly!

One day in early 2018, we had a conversation about him writing a book about his and our experiences. Since I knew he had read more than 80 books and hundreds of articles on the subject, I asked him if anyone had written about the relationship between geomagnetism and sasquatch. I saw a spark in his eyes when he said that, to his knowledge, such a book had not been written. I said, "Well, there you go!" And so, the idea was born, germinated for a while, and then he was off and running. Five months later, this book was in the editor's hands.

I believe his story is an important contribution to the growing body of knowledge related to the anomalous phenomenon known as *Sasquatch*. This book promises to intrigue novices and enthusiasts alike. So, strap on your seatbelt and enjoy the read!

—Julie Navarre, LMSW

G. Roger Stair

Forward

[1] "Quetico." Ontario Parks. Accessed January 02, 2019. https://www.ontarioparks.com/park/quetico.

Author's Preface

"Whether you think you can, or think you can't, you're right."

— Henry Ford

What this book is NOT about:

- ✓ This book is **NOT** an attempt to *prove* or *disprove* the existence of the so-called sasquatch.
- ✓ This book is **NOT** an attempt to engage in the ongoing and oftentimes vitriolic debate about *what* the so-called sasquatch is or is not.
- ✓ This book is **NOT** about *right* vs. *wrong* of the nearly overwhelming amount of conflicting—and sometimes confrontational—material about the so-called sasquatch as it exists in today's market place.
- ✓ This book is **NOT** about defending my beliefs about the so-called sasquatch phenomenon.

What this book IS about:

- ✓ This book **IS** about 17+ years of quiet, private, and revealing research into the so-called sasquatch phenomenon.
- ✓ This book **IS** about Earth magnetics and heart-head research and the way my early and exciting heart-head research unexpectedly transitioned to 17+ years of sasquatch research.
- ✓ This book **IS** about how I incorporated heart-head coherence into my research, and the amazing interactions with sasquatches that came from using it.

Chapter 1 presents our first encounter with a sasquatch. Chapter 2 shares how we used magnetic measuring devices and biofeedback units to figure out how changing Earth magnetics affects the human heart/head connection. Chapter 3 offers brief insights into the early years of errors and foolish decisions. Chapter 4 looks at the August 2011 research regimen, which followed the mistakes of our inexperienced early research routines; Chapter 5 discusses the September 2011 trip.

Chapter 6 tells about the July 2012 wilderness reconnaissance expedition, which turned out to be super beneficial; Chapter 7 covers the August 2012 trip; and Chapter 8 describes the October 2012 outing. Chapter 9 presents information from three expeditions taken in 2013. Chapter 10 looks at the year 2014, where the ideas of *habituation* and *familiarization* carved a road sign that directed me into a different and vital change in my research patterns. Chapter 11 offers detail on the 2015-2016 expeditions. Chapters 12 and 13 present specifics of the amazing and unbelievable 2017 expedition, which lasted for nearly six weeks. The year 2017 solidified the concepts of habituation and familiarization, along with their necessity in my continued research.

Chapter 14 discusses magnetism and geomagnetism, with a blurb or two on why we believe a clear connection exists between geomagnetism and sasquatches. Chapter 15 delves into some of the science that is so crucial to our work. Chapter 16 presents a discussion of dedicated sasquatch researcher Pearl J. Prihoda's theory related to Photosynthetic Piezoelectric Induced Transparency (PPIT).

Chapter 17 presents suppositions in lieu of conclusions, and Chapter 18 discusses plans and hopes for research in 2018 and beyond. Finally, Chapter 19 details some of my personal beliefs about sasquatches.

I have also included a selected bibliography. Lists of photos, diagrams, and tables follow Chapter 19. An index was included because when I read other's works, I find an index enormously helpful. I use endnotes for citation, and these appear at the end of each chapter for ease of access.

Again, my goal with this book is simple: Share my research findings and experiences with the so-called sasquatch species over the past 17+

years. To inform you upfront, my research approach did **not** include speaking to locals in the areas we conducted investigation; therefore, I've no tales to tell about local's beliefs or encounters. Rather, the purpose of my research was simply to engage with sasquatches in a manner in keeping with my earlier heart/head research. That is, I wanted to learn as much as I could about the sasquatch species while measuring and recording changing Earth magnetism variations during each sasquatch encounter. Secondarily, I chose to practice heart-head exercises while interacting with sasquatches. In the end, my daily practice of heart-head exercises that promote something called **coherence** is crucial to this story.

In terms of common conventions used throughout the book, I capitalize planet names when they appear as a proper noun; otherwise, they are lowercased. With minor exceptions, I do not use the word *Bigfoot*. Instead, I use the word *sasquatch*, which pluralizes nicely as sasquatches.

In this regard, I adhere to the convention of not capitalizing the word bigfoot that Thom Powell describes in *The Locals: A Contemporary Investigation of the Bigfoot/Sasquatch Phenomenon*. For instance, Powell wrote:

> Incorrect capitalization of the noun "bigfoot" leads us to assume that Bigfoot is an individual, not a member of a larger population. With one incorrect keystroke then, we imply that all sightings and track finds continent-wide are the work of a single, rather busy and well-traveled creature whom the tabloids call "Bigfoot" and who, like Elvis, pops up in some pretty strange places. An instant impossibility.[2]

As an aside, Powell is a retired middle-school science teacher and has authored other books, including *Edges of Science*, in which he wrote, "controlled experiments cannot be performed upon rogue entities that are intelligently directed." I say, "No kidding, Thom!" His novel, *Shady Neighbors*, shares a delightful story about children, family, and science in a unique yet believable way.

Powell is a decades-long sasquatch researcher and investigator who writes with the insight of one who knows the wilderness. He adroitly weaves mainstream science into most of his stories as only a teacher of science could do well. Everything he writes also has the taste and flavor of high wit—intermixed with droll humor. His books are well worth reading, I say.

Following our first unexpected sighting of a sasquatch, the team had breakfast out the following morning. To avoid folks learning about our experience with the sasquatch, I coined the name "Woody" so those who might overhear us while we were in a public realm did not know we were discussing the sasquatch phenomenon. I share this information because after almost two decades of using this name, it feels damn awkward to use the word *sasquatch*. Even so, rather than confound the sasquatch world with yet another unneeded name, I will not use the name *Woody* again in this book; rather, I use the term sasquatch.

Finally, beginning with Chapter 6, I offer proposals to aid you in following my thoughts as I move from knowing absolutely *nothing* about sasquatches to knowing a tiny bit less than nothing as our research unfolds. Hopefully, the proposals succeed in their task.

---

[2] Powell, Thom. *The Locals: A Contemporary Investigation of the Bigfoot/Sasquatch Phenomenon*. Surrey, B.C.: Hancock House, 2003, 6.

Prologue

"An idea, like a ghost, must be spoken to a little before it will explain itself."

— Charles Dickens

I walked into the controversial field of sasquatch lore strictly by accident, and unfortunately, with nose wide open. Granted, I had read stories and articles of a "hairy, human-like ape," and found the idea of sasquatch quite fascinating, albeit not caring if the stories were true because they obviously did not relate to my primary research interests. Therefore, I did not pay excessive attention to sasquatch or the myriad stories and hype surrounding them. I lived in a safe, cocoon-like bubble of sasquatch innocence, with no personal mindfulness of hairy, human-like apes.

Again, I had my research to consider—research that, although unrelated to sasquatch in any known fashion, led me literally down an odoriferous wilderness trail smack into a life-altering decision quandary. Wait! I'm getting a bit ahead of the story. Let me backtrack a bit before plunging into this unexpected, yet captivating saga.

The Path to Sasquatch and Me

I founded my consulting business, Plan Your Future Now (PYFN) in 1989 as a Doing Business As (DBA) in the state of Michigan. In 2001, I filed as a Limited Liability Company (LLC) and became Plan Your Future Now, LLC.

Among other areas, PYFN has great interest in Earth magnetism and how changing Earth magnetism affects the human body, mind and, if valid, spirit. Our interest in the way changing Earth magnetism affects the human heart/head connection led us into seemingly unlikely places. Initially, this included many wilderness areas within the western Upper Peninsula of the state of Michigan, largely due to friends who had a lodge in the area telling us fascinating stories about changing magnetism and unusual phenomenon occurring in the broad expanse of forest within which they chose to build their lodge.

Due to the innate nature of Earth's changing magnetism, an ongoing literature review became the norm. For instance, I had to become familiar with different themes within each designated research locale because each area played both an individual and a cumulative role in the way Earth magnetism changes affect the human body.

To aid our research and ability to understand zones of interest, we secured countless maps. The most important *initial* map for our changing Earth magnetics research was the *Total Intensity Aeromagnetic Map of the Western Part of the Upper Peninsula Michigan.*

This map was key to our research relating to Earth magnetic changes because it provided us with a verified, known baseline of typical earth magnetic values for the western Upper Peninsula of Michigan. Armed with this information, we checked for Earth magnetic changes using a variety of instruments. Without this map we would have had to spend hundreds of hours and thousands of dollars creating magnetic baseline data, which most likely would not have happened.

Five years of research into the heart/head connection interacting with changing Earth magnetics produced phenomenal results. For instance, data showed *changing Earth magnetics affects the human heart/head connection in amazing ways.* Because sasquatches are carbon-based creatures—much like humans—I propose the same applies to them. This idea takes root here, is nurtured as the story unfolds, and finally blossoms into full-fledged premises near the end of the book.

Furthermore, evidence presented throughout the book points to a nearly unbelievable conclusion: *sasquatches can manipulate Earth energies for specific results.* This is a bold, fascinating, and controversial proposition, yet years of research data strongly support this daring statement. Wait and see.

Please do continue reading to the end. This alone will allow you to formulate an informed opinion about this audacious statement and the many incredible and seemingly far-fetched experiences detailed throughout the book. In short, this research has been an amazing journey into the enticing realm of sasquatch, and we are pleased to bring it to you in this format.

After receiving the manuscript back from the editor, I had serious internal struggles over the idea of publishing it as a book. These struggles occurred over several weeks; in the end, I obviously resolved those dilemmas because you are reading this! Hope you enjoy it!

Chapter 1

The Beginning: Fall 1999

"The beginning is the most important part of the work."

— Plato, The Republic

The year was 1999, mid-September, mostly sunny, in the upper 60s, and with a slight breeze—about five miles per hour. Leaves of the deciduous trees were beginning to succumb to the natural cyclic progression of fall, many turning bright red or shiny gold. These brightly hued colors were sporadically interspersed with those reticent to turn color yet, and these shimmered through as optimistic green.

We were wrapping up yet another excursion into the wilderness to chart and document how the heart/head connection alters during changing Earth magnetics. For the most part, we were pleased with our results and took time to enjoy the beauty of a fall day hike. We had embarked upon the hike in this lovely wilderness and our conversation was intermingled with upbeat feelings about a mostly successful data-gathering trip.

There were five of us on this hike into the wilderness: my wife Maggie, a dear friend who called herself BoB, a male friend I've nicknamed Brushtracker, his wife, who I nicknamed Dragonfly, and me.

We had just climbed a small hill north of a narrow, vibrant, stream named Bluff Creek in Iron County, Mich.

Bluff Creek, Michigan ... Not Northern California

At the time of this hike, none of us were aware of the irony of the creek's name in relationship with sasquatches. That is, we had not yet consciously drawn an association between that famous, controversial American film—known as the Patterson-Gimlin film, and the name *Bluff Creek*. This film depicted a sasquatch walking along a stream in northern California named Bluff Creek.

Again, we were in Iron County in the western Upper Peninsula of Mich., and the sun was shining brightly from a penetratingly deep blue sky. At times, the sun's rays reflected through fleetingly dancing, puffy white cumulus clouds. These entwined with blending gray clouds, as if teasing us with a gentle reminder to enjoy the beauty of the fall colors before the crushing onslaught of winter hit and buried the area under the weight of tons of snow.

We turned east onto an old, overgrown logging road while moving deeper into the wilderness. As I recall, we were commenting softly about the beauty of the wilderness while walking in a relatively tight group.

Of an immediate sudden, we simultaneously stopped moving, ceased speaking, and gawked at one another with befuddled wonder. I suspect we were each having a similar thought: "What in the world is that *putrid* stench?"

PHEWH! Yuck! Faces covered with hands, clothing, or both, we looked around us in all directions, yet we saw nothing.

As our minds began to process the wretched aroma, we heard a distinct movement due north of us; our best guesstimate was 15-20 yards directly north of our position. Brush crackled, twigs snapped, and it sounded like a large draft horse was moving rapidly through the thick brush and timber in a north-northwest direction. How could that be? It was seriously dense in there.

We remained frozen in place for a brief period—maybe 30 seconds, maybe 90 seconds, certainly no longer than that. As the noise of movement faded away, the putrid, decaying, foul odor also diminished—but only slightly.

Unbeknownst to us at the time, we had just been introduced to an alleged sasquatch who apparently was telling us to vacate its territory. Ostensibly, not a very friendly first introduction!

By this time, we began jabbering like a roomful of pre-school children just prior to being overwhelmed with myriad goodies on Halloween day. Of course, none of us really heard what the others were saying for the first few seconds of this jabber-fest; we were each experiencing our own inner tumultuous response to a rather unexpected and extremely malodourous and anomalous phenomenon. Clearly, if one of us was the perpetrator of that awful disgusting odor, someone was extremely ill. Mercy!

Reason ensued after a few seconds of the jabber-fest—at least momentarily. We continued to hear SOMETHING move in a north-northwest direction, although we saw nothing. Brushtracker decided to track whatever it was through the dense brush and foliage.

The rest of us chose to stay on the trail and follow the smell, thereby avoiding the seeming impenetrable thickness of the brush. I mean, who wants to fight dense brush and thorn-induced scratches when a practically open trail was available? Although the stench of the original odor had subsided somewhat, its lingering sensation was strong enough to easily follow, so follow it we did.

We walked the overgrown logging trail west back to the main trail-road intersection from Bluff Creek. At that point, it is a longstanding dirt road named Old 45, giving access to Bluff Creek for those who wish to fish or simply dilly-dally along its banks. There were no trails along the creek, so anyone wishing to amble parallel to the creek needed to battle the woods and dense underbrush. Most folks not fishing dawdled along the open gravelly banks fronting the old road while looking for cool quartz stones. A memory, if you will, of the lovely area.

No one else was there that day. Also, Old 45 was barricaded off on both the southern and northern sections at the top of the ravine through which flows Bluff Creek. A bridge once spanned this deep ravine; however, it had been removed many years prior to our visits.

Four of us slowly walked due north on Old 45 for about 500 feet before noticing the intense odor coming from the thick brush-infested wilderness to our left, and it seemed to plot directly west. We gave no thought to looking for crossing footprints. Even if we had, probably none were visible; Old 45 was old, graveled, well-worn, and much too hard to easily show prints.

We waited patiently for Brushtracker. After approximately 20 minutes, he appeared on Old 45 about 30 feet from where we were standing. Dang! He did a superb job tracking the smelly and noisy movement, although it took him a bit longer to arrive at the same location we did. Not saying we giggled about this or his slightly disheveled state; even so, we *might* have had a *little* fun at his expense.

Once we settled down, we decided to pursue the smell as it progressed in a west-northwesterly direction. Approximately 100 yards into the woods from Old 45, we saw three huge old white pine trees growing in a large triangle pattern. Two of us hugging the trunk could not wrap our arms around them. We dubbed them the "three sisters" for later observation and locale purposes, knowing we would return to them at some point.

We moved west-northwest while following the diminishing odor. I and Brushtracker decided to move smartly ahead of Maggie, Dragonfly, and BoB. Brushtracker and I rapidly covered ground while following the odor—less intense though it was—as we moved more deeply into the wilderness. The less intense odor stoked our desire to keep going as we were following only the odor at that time and didn't want to lose it.

We did begin hearing movement noise, along with a bit stronger odor, once we had walked about 300 yards farther into the wilderness. We gave no thought at that time to the way the odor weakened and then strengthened as we moved along. At that point, though, I removed my camera from its case and readied it for picture snapping. As we continued our rapid pursuit of the odorous noise movement, we viewed a slight break in the forest just ahead of us.

It wasn't really a meadow; rather, an unexpected open space where the trees thinned out and the brush transmuted into bracken ferns interspersed with white baneberry, and bergamot—a lovely, but poisonous, lavender flower with pointy green leaves. This relatively small oasis space was surrounded by a forest of balsam fir, red pine, white pine, jack pine, white spruce, quaking aspen, paper birch, sugar maple, and black ash. Truly a lovely and eye-catching area!

While it was difficult to see directly through the forest into the small open space, after a few more steps we found a spot that did allow a clear view ahead. We stopped, looked at each other, and both exclaimed, "What!?"

At the northern edge of the open space we saw an **enormous** bipedal anomaly. Holy WOW… have mercy! This startled the two of us into an abrupt silence, even more so than the earlier onslaught of stench did, if that is even possible. The bipedal object was standing fully upright, had hair covering most of its body, and was facing away from us as it moved calmly toward the forest on the north end of the open space. We could see its back, legs, and long swinging arms. It seemed completely covered with hair, mostly brown, although I saw a bit of a dark blonde streak along the back of the arms and legs. Do these creatures have hair stylists? Anyway, we guessed the hair length at about four inches.

Just prior to stepping into branches that the forest seemed to open for it, it moved its shoulders to the left as it turned to fleetingly glance back at us. We were too far away to see specifics on its face; however, it appeared to have no (or nearly no) hair on its face and very little, if any, neck. Its eyes left us with the impression they were large and deeply set, and the face itself looked rather dark from our position. This assessment is probable without being definitive because we did not see the face as clearly as the back side of it. We later decided its face looked like deep-tanned leather. I'm sure you know how *later* conversations (often known as Monday night quarterbacking) tend to fill in missing details of an otherwise inexplicable or highly filled adrenaline experience.

At any rate, the top of its head seemed to be somewhat conical shaped, although not grossly pointed. Its shoulders were like "wow broad," its arms long and hanging about to its knees, and clearly, whatever the heck this thing was, it was strong. Damn! This THING was muscular. We saw no clothing of any type on any part of its body and, while there was no way to confirm it, we guessed its height to be between 7½ to 8 feet tall. It was unquestionably much taller than us, stronger than us, and did not seem at all frightened that we saw it. Nor were we frightened of it as we watched it slowly fade into the forest. We were simply too astonished and taken aback to be frightened. Or, perhaps, too foolish at that moment? Hmm!

We stood together stunned, having no known reference in our experience with which to compare it, other than literature we had read that discussed something known as sasquatch. Surely, this couldn't be THAT! Of course, as astonished as I was, I never gave a thought to putting my camera to eye and snapping pictures. Shame on me! Rather, we essentially glanced at each other with what I'm quite sure were blank and mindless looks on our faces, each wondering what in the world we had just seen, and not knowing how to discuss it. Although the experience seemed to last forever, it took no longer than a few seconds.

We quickly turned our heads to watch the entity as it stepped blithely into the reaching arms of branches. Say what? I would have sworn those branches *opened* for the bipedal anomaly's entrance into the green, red, and gold leaf-colored forest. Dang, Dorothy! Is there really an Oz? Suddenly, we realized we no longer smelled the incredibly annoying stench we had been following and were clueless as to why it had disappeared.

Then, as we watched the creature fade quietly into the forest as if it were merely disappearing, it made a subtle sound. I've no real recollection of how the sound struck me back then, only that the sound was neither hostile nor frightening. Maybe that sucker was chuckling about our blank, mindless looks and thinking "Damn dumb humans!" Whatever!

After what seemed a lifetime of being stunned (probably about 2-3 minutes), Brushtracker and I tried to each relate to the other what we had seen and experienced. Darn it! Another temporary uncontrolled jabber-fest. Okay … deep breath and try again. We listened to each other's sense of what we saw, recognizing we had each perceived the thing in a very similar way. After this, we shrugged and decided to say it must have been a so-called sasquatch although merely saying the word felt awkward, even uncomfortable as it stumbled and twisted out our mouths past nearly numb lips.

Now then, to catch you up on the extent of our knowledge of sasquatch at that time, it was basically minimal to naught. We had each read a bit about it from time-to-time, yet neither had deeply delved into its history; nor had we tried to keep up-to-date on what was being written about it. So … there we were; we had just had an *anomalous experience* with something we really couldn't name or speak intelligently about, yet we were totally happy to share our ignorance with each other. Oh well, friends obviously have some use, eh? As an aside, the phrase *anomalous experience* blossoms into a crucial discussion later in the book.

We backtracked toward the other three members of our cluster of five. They had continued toward us, albeit slowly, so it didn't take long to regroup. We, of course, excitedly shared our experience. Guess we expected them to believe us. How foolish! They were convinced we were jerking their chains, so they chortled, chuckled, and outright laughed as we strove to make them understand we *really* did see something big and hairy!

We had cheese, crackers, and water in our backpacks, so we decided to take a bit of a break. We sat on the forest floor, kneeled, or stood around discussing our situation and then BoB noticed a small plane flying overhead. It was flying low and, considering the rapid turning of leaf colors, we assumed its inhabitants might be appreciating nature's splendorous beauty from above.

Hmm! A bit later, we noticed it flying overhead again; at that time, though, we heard what can only be described as a type of low whistle—followed at once by movement in the forest. Say what? We all looked at each other and then began laughing. Laughing? Okay. Let the effects of *anomalous experience* reign, thereby creating a false sense of momentary euphoria.

At least now, though, the women were beginning to believe that we *may* have actually seen something; that we were not simply fabricating our story. We packed up our snacks and began wandering somewhat aimlessly while deciding what next to do. The plane flew over again and, once again, we heard a low whistle and movement in the forest, albeit farther away than before. We chose to follow the whistling sounds for a bit and were very thankful the stench had disappeared.

It seemed as if the THING we had seen chose to circle around us yet stay completely out of sight. Not sure how it moved so quickly. Comparing our size to it, though, because that THING was indeed enormous, and with long legs, Brushtracker and I were convinced it would surely move much faster than us. Clever boys, eh?

As the five of us walked through the forest, this time in a south-southwest direction, the plane once again flew overhead. And once again, we heard movement in the forest at the same time we heard the low whistle. BoB wondered if there could be a connection between the plane flying overhead and the noises we heard in the forest. Good insight, although none of us could logically tie the two together. We felt, instead, that the low whistle and movement related to our presence in the forest. Seriously, what's wrong with a bit of conceit? At this point, I suspect it's possible someone in the plane was trying to view the same THING we were experiencing. We'll never know.

By now, though, the sun had begun to drift gently down the horizon, and it's best to be out of the forest before dark unless you are fully prepared for an overnight stay. We were not; ergo, we headed out. As we were leaving, we once more heard the low whistle and movement in the forest as whatever it was pressed onward in the opposite direction from us.

We arrived at our vehicle just before dark and headed back to the lodge. This seemed like enough idiocy for one day! Unless you include that we were up nearly the entire night repeatedly rehashing the events of this highly unusual day. Not surprising, eh?

Chapter 2

Heart/Head Research and Earth Magnetics — 1994-1999

"Earth's polarity is not a constant ... the flow of liquid iron in Earth's core creates electric currents, which in turn create the magnetic field."

— NASA, November 2011

This decades-long sasquatch research project began with the strangely unusual encounter in 1999, and my avid interest in pursuing it grew over the years. As such, I did not enter this research endeavor with a specific hypothesis in mind. Rather, bumbling through different research patterns eventually led me down a path of ongoing realizations about sasquatches, their environment, and our later interactions with them.

One such realization, although instinctive at first, turned into a mainstay for ongoing research. That is, I deduced early on that my heart-head research involving Earth magnetism and the human body likely also played a role with sasquatches. Why? First, because both humans and sasquatches are carbon-based lifeforms. In addition, they have similar anatomical constructs and they coexist within the magnetic-gravitic makeup of Earth. Second, the path the first sasquatch we smelled and saw followed a similar changing Earth magnetic line we had used during our heart-head research. Hence, I deduced that changing Earth magnetism might also affect sasquatches. More on this later. This realization, though, did raise a couple of early questions:

1. How would I go about trying to prove Earth magnetics affect humans *and* sasquatches in similar ways?
2. How would unknown variants between human and sasquatch anatomies cause differences in reactions to changing Earth magnetism between humans and sasquatches?

Clearly, I was entering a realm of unknowns and, I didn't know yet what the unknowns were; only that I had to move forward. I did this, albeit somewhat blindly at the start, and the many lessons surrounding changing Earth magnetism and sasquatches began!

Over time, I absorbed numerous lessons, and finally stepped into directed sasquatch research efforts. By documenting all my efforts—from sorely misguided and successful—I identified intriguing patterns over the years. It didn't take long during the early stages to realize magnetic and geomagnetic correlations with sasquatches seemed a prominent reality, so I pursued that avenue with vigor. It wasn't until 2011 that I became aware of the ideas of habituation and familiarity. At that time, I began to intentionally include them in my research. From the beginning, though, I believed in and, used, deliberate *coherence* to lay the groundwork for intentional positive interactions with sasquatches. Because the concept of coherence is crucial to this story, I elaborate on it throughout the rest of the book.

Confusing Results?

It's time to step back now, though, and take a peek at the way early research results showing effects of changing Earth magnetism on the heart/head connection led me eventually down a road of education, wonder, challenges, and a life-changing decision. *Education* came into play in the way I learned to apply early magnetism measurement skills differently while researching sasquatches. *Wonder*, because to this day I am decidedly beguiled with the beauty, range, and power of magnetism. *Challenges*, because I had to reeducate myself as I moved—inadvertent though it was at the beginning—into the invigorating chronicles of sasquatch and me. *Life-changing*, because from the moment I smelled and saw my first sasquatch, I was captivated by the sasquatch phenomenon in a way that boggled my mind and touched my soul.

Speaking of magnetics and magnetism, our first and intriguing introduction to the so-called sasquatch was not the only unusual anomaly we experienced that year. The year 1999 also brought unexpected changes in our heart/head magnetic data. For instance, I was astounded to discover that known patterns of changing Earth magnetism mapped during earlier years of research *no longer held true*, and I had no logical reason at that time to describe why that was so. That awareness came later; hence, I was a tad or so befuddled once I realized we were recording strangely unfamiliar and disparate changes in measurements from previous years.

It wasn't until the winter of 1999-2000 that ongoing extensive literature research overturned a rock of confusion for me to expose a sense of temporary clarity rather than grubs, ants, worms, or garter snakes. One would think clarity is good; unless, of course, newfound clarity opens one to something that exposes a previously unknown phenomenon. Oh, well; such flow the myriad unpredictable currents of research.

Between 1994-1999, our team drove thousands of miles across the western Upper Peninsula of Michigan. We traveled repeatedly between the Keweenaw Peninsula—which protrudes into Lake Superior at the northern tip of the Upper Peninsula—south to the northern boundary of Wisconsin. From Iron River, Mich., we traveled northwest to Ontonagon, Mich., where Lake Superior waves resoundingly splash along the far western Upper Peninsula, and the wide mouth of the powerful and impressive Ontonagon River merges with the deep, cold, and sometimes wildly tumultuous waters of Lake Superior. Miles south of Ontonagon, Mich., is an area of deep wilderness known as the Porcupine Mountains. While not mountains in the *Rocky Mountains* sense,

Porcupine Mountains has an amazingly gorgeous scenic overlook. Sasquatch notwithstanding, well worth a look-see—if you don't mind a bit of an uphill hike.

Aside from the fact that Michigan's western Upper Peninsula is a beatific unfoldment of nature that we enjoy immensely, our travels uncovered many areas of focus where changing Earth magnetics were readily measurable. Thus, it became a matter of choice as to where we would begin our early changing Earth magnetics research.

We were quite fortunate in the first stages of research to avail ourselves of free lodging. Close friends of ours had built a log lodge on the shore of Tamarack Lake, 20 or so miles west and a bit north of Iron River, Mich., and a few miles north and a bit east of Watersmeet, Mich. Both Iron River and Watersmeet reside on Highway 2, the main thoroughfare (mostly a two-lane highway) stretching across the southern section of the Upper Peninsula, dipping momentarily into Wisconsin then back into Michigan as it continues to Minnesota.

Jay Leno and the Watersmeet Nimrods

A dynamic aside: Watersmeet is home to the *Nimrods*, the local high school sports team and mascot featured on ESPN commercials circa 2004. The Nimrods also had a lively appearance on the *Tonight Show* with host Jay Leno prior to his retirement. Temporary fame washed over Watersmeet, due to the mascot name—an unusual mixture of biblical and warrior origins—which netted the school district about $500,000.00, funds that aid high school graduates wishing to attend college. Neat! Especially for a small community with a population of a little over 1,400 souls. Go Nimrods!

Our lodge friends unlatched their hearts and doors to my wife and I, complete with an open invitation to bring others with us. In brief, my wife and I did take many others to the lodge over a period of years, and those "others" turned into excellent subjects for research. Yay! Of course, those poor subjects were able to enjoy the beautiful lodge on Tamarack Lake for free, the serenity of long nature hikes, long drives through the spectacular flora of nature, the excitement of oftentimes strenuous forays into the depths of the wilderness at literally all times of day and night, and the endearing companionship such efforts entail. Win … win!

Sadly, my log lodge friend has passed. A few years following his death, his wife, a charming, delightful, open-hearted person who has no equal, sold the log lodge when its upkeep became too much. She is a person to whom we owe an unpayable debt of gratitude for, without her love and support, this book would still be in an ethereal realm of a fleeting idea rather than moving into the physical world as an enchanting story of sasquatch and me. She is as dear a friend as one could imagine, and undeniably stalwart and loyal in her friendship.

Now then, back to our research beginnings. At the time, common sense suggested that we plot our research locations within driving distance of the lodge. Fortunately, we had discovered many locations fitting this criterion. Following a diligent poring over maps combined with lengthy consideration of logistics, wildlife demographics, and the geography associated with changing Earth magnetics, we chose to use the Bluff Creek area mentioned in Chapter 1 to start our heart-head data gathering in 1994.

Again, I was unaware at the time about the irony of the name Bluff Creek and its association with the 1967 Patterson-Gimlin film, an American motion picture of an unidentified subject which the filmmakers said was a *bigfoot.* This famous and controversial American film was filmed alongside Bluff Creek in Northern California.

Please know this is not to imply I am slow; rather, merely to say I was a sasquatch innocent, a bit ignorant about sasquatches, or perhaps, a combination of both. Eventually, though, both my sasquatch innocence and ignorance were assuaged through extensive literature reviews, online research, and a newly adopted ongoing realization that so-called reality is not always what we are taught it to be.

1999 Heart/Head Research Data Summary

Several years of data patterns altered substantively in 1999 and at the time we had no idea why this occurred. First, the data patterns gathered between 1994-1998 disclosed that changing Earth magnetics affect humans in unusual ways. This is astonishing, to say the least. Second, the data gathered during the first few years related to blood pressure, pulse rate, pulse oximetry, and basic biofeedback to measure something known as heart rate variability (HRV). Third, early recorded magnetic data presented as change in a site's magnetism via use of a Null Meter we borrowed from a kind-hearted Michigan high school physics teacher. While slightly constricted due to our instrumentation, it's quite tantalizing how much verifiable data can be gathered with what many might consider inadequate equipment. Inadequate? I think not, for we learned early on how to accurately use basic equipment to produce startling results.

Help from An Angel

I received unexpected outside help to continue the heart-head research. An angel surfaced during our early work and, through her open-hearted encouragement and generosity, she helped us invest in field instruments that allowed us to continue as well as expand our field research efforts. Armed with upgraded instruments and equipment, we plunged enthusiastically into ongoing heart-head data gathering.

It goes without saying that, minus her financial support—as well as her ongoing encouragement—this work would not have unfolded as it has. Nor would this book. One minor note: this angel has no clue how our initial research moved laterally, albeit unintentionally, into this story of sasquatch and me. Surprise!

As mentioned earlier, we have many years of gathered heart-head data. The gathering of this data served to get us into our research areas and produce verifiable, reproducible results detailing clear and definite associations between changing Earth magnetics and humans, particularly within the heart rate variability and autonomic nervous system zones.

I choose, however, to not take up space or time displaying early accumulated heart-head data here; rather, I prefer to keep focused on the gist of this book, which is the ever-elusive sasquatch and me. Later, I discuss the importance of heart rate variability (HRV), the autonomic nervous system, and a phenomenon known as coherence. Please do continue reading, for coherence and HRV indeed played a crucial role in sasquatch research strategies and tactics—and continue to do so today.

Drastic Changes in Our 1999 Field Research Data

We had mapped various areas where changing Earth magnetics occurred consistently. This had become a given for us because these areas produced similar data at similar times of days for four years prior to 1999. Each time we were in those locations, we produced irrefutable evidence to prove a clear connection between changing Earth magnetics and humans. In fact, anyone using the same instrumentation in the same areas at the same times would have produced similar or the same results. Succinctly, our field research was undeniably verifiable as well as reproducible. Got to love replicable research.

While in the field in 1999, we were not able to figure out why our research in these specific locations changed or why they changed at the time they did. Merely that results did indeed change, and they did so unexpectedly. Fortunately, questing literature reviews opened my eyes and mind to a previously unknown reality associated with that area—that of extremely low frequency (ELF) communication. ELF plays an important role in my ongoing sasquatch research, so please do remember the acronym.

Much to my surprise, and perhaps a bit of chagrin, I learned that ELF had played a significant role in our ongoing heart-head field research. Momentarily, I'll explain how the U.S. Navy's use of extremely low frequency near the areas we conducted field research affected our data, and how changes to the ELF project in a two state area in 1999 smashed an unforeseen hammer on the data we typically gathered.

Interesting stuff, ELF. Powerful stuff, ELF. Controversial stuff, ELF. Finally, magnetic stuff, ELF. To understand how ELF affected our data gathering, it is crucial to first glean a sense of how ELF works. I'll keep this brief and simple.

What the ELF?

Upon discovery of this phenomenon, I felt somewhat disheartened. After learning more about ELF, the geology of our chosen area of research, and how the qualities of electricity and magnetism drives the power of ELF, that disheartened feeling flipped to amazement.

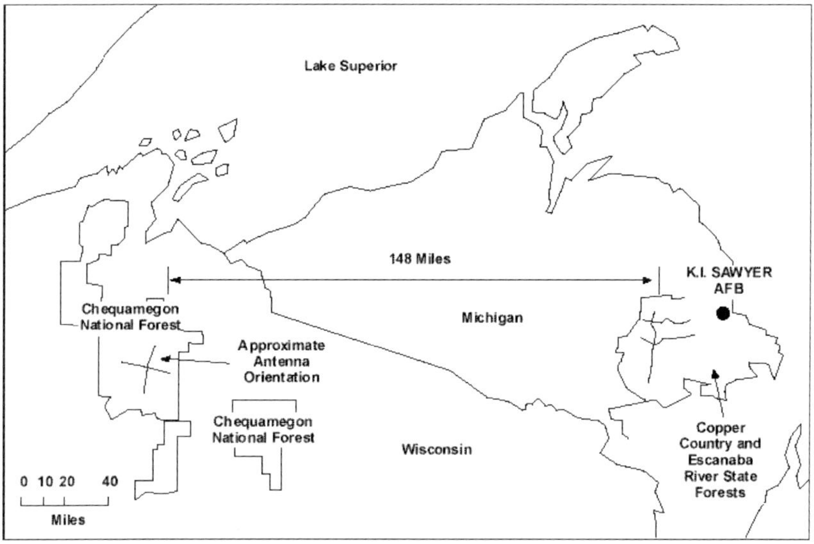

Diagram 2.1: ELF transmitter facilities in MI and WI

Diagram 2.1 shows two ELF locations[3]—one in Republic, Mich., and one in Clam Lake, Wisc.—as they existed during our early field research. Diagram 2.1 also shows a 148-mile linear indicator between the two facilities. The distance between the facilities was a plus to ELF, not a detractor—and these ELF operations caused interference with our field magnetic measurements.

While looking at diagram 2.1, also take note of the *areas* in which the ELF facilities were built. For example, the Republic facility was built within the Escanaba State Forest, while the Clam Lake facility was constructed within the Chequamegon National Forest. The significance of these locations is simple. According to the U.S. Navy, the locations were selected for minimal population. Furthermore, constructing facilities in forests included clearing lanes of land for antenna transmission lines. The Navy believed doing that would help forest wildlife by creating paths of open areas for it to feed and travel. Eventually, the Navy worked with the Wisconsin Department of Natural Resources (WI DNR) to introduce elk into the Clam Lake area.

The Navy had conducted studies on health effects of ELF on humans and had decided that the frequencies in play were not harmful to humans. Much later, this thought was challenged by local populations; however, the Navy continued to support the idea that ELF signals did not harm humans or animals. Regardless of the controversies surrounding ELF and local health issues that arose over the years—some of which continue to this date—extremely low frequencies assuredly affected our heart-head research field data. Myriad data support this statement.

The ELF facilities, dubbed "Project ELF" were the brainchild of the U.S. Navy and came into being to aid with deep sea submarine communications. Prior to that, and as shocking as it might seem, the 1960s saw Navy officials pursuing a highly ambitious ELF plan named "Project Sanguine." With clear intent, the Navy "wanted to bury a gigantic grid of cables under roughly *41 percent* of the state of Wisconsin in order to turn its bedrock into the world's largest radio antenna."[4] Following the failure of Project Sanguine, the two smaller facilities in Republic and Clam Lake were built.

The ELF facilities, while separated physically by 148 miles, functioned with synchronous operation due primarily to the bedrock geology of the entire area encompassing both the western Upper Peninsula of Michigan and northern Wisconsin. In short, the ELF facilities in Republic and Clam Lake were designed to make practical use of the geologic formation known as the Laurentian Shield—also called the Canadian Shield. This geologic bedrock formation offers low electrical conductivity—a requirement for ELF antenna efficacy. In effect, it was the natural geologic construct of the Laurentian Shield that allowed the two ELF facilities to function as a "synchronous two-site operation with minimum signal degradation."[5]

Control of these facilities fell to the K.I. Sawyer Air Force Base in Marquette County of Michigan's Upper Peninsula. The K.I. Sawyer base officially closed in 1999; this suggests that changes in the Project ELF operation occurred then as well. As a point of interest, in today's world, part of the earlier Air Force base's 5,200-acre plot serves as Sawyer International Airport, with bragging rights for having one of the longest runways in North America. Quite a feat for a far northern Michigan county with a population estimate of 66,502 as of July 01, 2017![6]

Basics on How ELF Works

Photos 2.1[7] and 2.2[8] show aerial views of the Clam Lake,[9] Wisc. and Republic,[10] Mich. facilities, respectively.

Photo 2.1: Aerial view of Clam Lake, WI ELF site

Photo 2.2: Aerial view of Republic, MI WLF site

A United States Navy Fact File sheet states, "Each ELF antenna works as an independent horizontal electric dipole [a pair of electric charges or magnetic poles, of equal magnitude but of opposite sign or polarity, separated by a small or large distance[11]]."[12] These two transmission sites orchestrated their transmissions, allowing coverage to most of Earth's oceans. Again, the ELF facilities were built to make practical geographical use of the innate nature of the Laurentian Shield bedrock layer. Go Earth!

Because the qualities of a dipole are key to the way ELF affected our data, allow me to further address the concept of a dipole as used in the earlier quote. Ronald Merrill, author of *Our Magnetic Earth*, wrote that a dipole "consists of plus and negative magnetic charges, [and] it both repels and attracts any other magnetic charge."[13] This is a key factor in ELF.

According to the U.S. Navy, the "geological formation [of the Laurentian Shield] channels ELF currents deep into the ground and effectively increases the size of the antenna for more efficient signal transmission. The conductivity of the bedrock layer helps to improve the efficiency of the antenna system."[14]

They continued with, "The areas chosen for the ELF system have low conductivity rock (rock that does not conduct electricity well) that produce the best results for creating an ELF antenna. In these areas, electrical current flows deep into the ground (hundreds of meters) before returning to the opposite antenna terminal ground."[15]

Photo 2.3 shows transmission lines from the Clam Lake facility. Note the similarity between these and regular power lines.

Photo 2.3: Antenna transmission lines at Clam Lake, Wisc. [16]

Now then, the way ELF works to send signals to submarines stationed around the world might seem complicated at first blush; however, premises describing the basic workings of an ELF system can be readily absorbed.

Next, let's look at diagram 2.2. It shows a schematic,[16] or a visual if you will, for the basics of how ELF communication works. Inspection of this schematic will aid in understanding how an ELF station functioned.

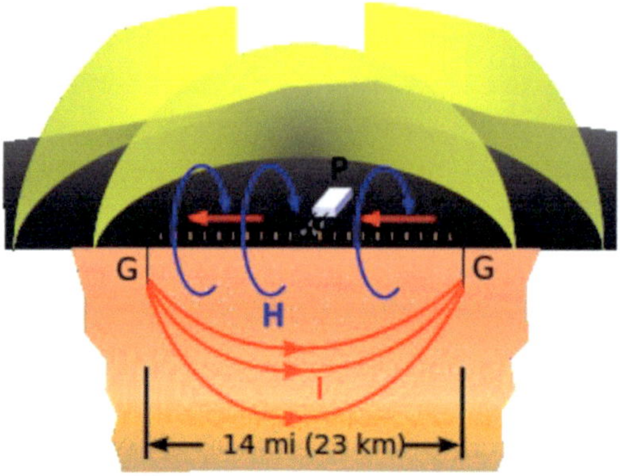

Diagram 2.2: Schematic of ELF antenna function[17]

The ELF project stations included "a transmitter, pole-mounted antenna cables, and buried ground terminals where the antenna currents enter the earth."[18] The actual transmission lines were mounted overhead on poles resembling ordinary power distribution lines as noted in Photo 2.3. The start and end of each transmission line was buried deeply within the Laurentian Shield; these served as the dipoles mentioned earlier.

An external power source provided alternating current (AC) to each end of the transmission lines—see the box labelled "P"—as shown in diagram 2.2. Note how the arrows depict AC current flowing in *one* direction only as it passes through the overhead transmission lines. Actual AC current flows in *both* directions in contrast to the indicator lines in the diagram. For our purposes, we need only be aware that an electrical circuit was completed between the two ELF stations via huge grounding poles buried deep underground. Most importantly, the completed electrical circuit made use of the natural bedrock geology of the Laurentian Shield in a way that unequivocally affected our heart-head research data.

The AC current flowing through the transmission lines induced, or, if you will, persuaded a corresponding magnetic field to exist. An induced magnetic field is seen represented with the lines labelled "H" in diagram 2.2. Note how the induced magnetic field appears to circle the current flowing through the transmission lines. In turn, this induced magnetic field created waves of energy as depicted with the half-circles existing above the arrows. I return to the idea of waves of energy and the innate nature of the Laurentian Shield later in the book because the concept of electromagnetic waves is vital to the overall story of sasquatch and me, not to mention the bold premises put forth later in the book.

To summarize, a power source fed AC current to the transmission lines. This current flowed through the transmission lines into deeply buried ground poles, which, in turn, served to form a complete electrical circuit between the two ELF stations. Current flowing through the completed electrical circuit induced a flow of magnetic energy around the transmission lines. This magnetic energy created waves of energy that both traveled through the Laurentian Shield as well as shot up into the ionosphere, which, in turn, traveled—happily, I presume—around the world.

After thousands of miles of travel, these waves of energy filtered down through the ionosphere, into and through the mesosphere, through the ozone layer existing at the top of the troposphere, and, finally, dove valiantly into the waters forming the oceans of the world. Here, the long-traveling ELF signal tickled a long wire that submarines trailed out the backside of their ships. The long-wire antennas trailing behind the submarines were designed specifically to pick up the ELF incoming signals. This was a very slow, yet efficient and silent method of communication with submarines.

Pretty cool, eh? That is, until we discovered how the placement and functioning of the ELF facilities affected the data we gathered. Nah! Still cool how the physical construct of Earth promotes such an amazing phenomenon. Particularly because, over time, I also discovered how this phenomenon affects sasquatch behavior. Hmm!

ELF's Penetrating Reach into Heart-Head Data Gathering Areas

I suspect you are now fitting together pieces of the ELF puzzle in a way that makes sense of how that naughty ELF stuff blew our 1999 data gathering singularly into the atmosphere above our heads and bedrock beneath our feet. From here, it is an easy step of logic to figure out why Project ELF affected our data, particularly our magnetic measurements, because—the magnetic waves created by ELF flowed freely throughout the entire Laurentian Shield!

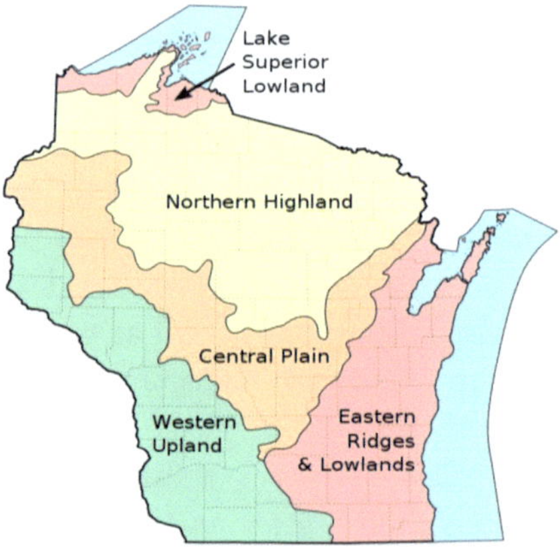

Diagram 2.3: Geographic areas — WikimediaCommons

This leads to new information; that is, according to Wikipedia, "Wisconsin is divided into five geographic areas."[19] Our area of sasquatch research interest lies in the Northern Highland geographic area, which includes the enormous 1,530,647 acres of the Chequamegon-Nicolet National Forest. The Laurentian Shield underlies this entire area, helping to make it a dynamic zone for successful sasquatch research. Diagram 2.3 shows a map of the five geographic areas of Wisconsin. Again, the huge Northern Highland geographic area is our focus of research.

Due to the geologic construct of the Laurentian Shield working with the Navy's directed ELF antenna currents—unbeknownst to us at the time—those electromagnetic energies flowed beneath our feet! Those energies, then, created amazing alterations in Earth's typical magnetic field, causing magnetic values to greatly fluctuate between those provided by the *Total Intensity Aeromagnetic Map of Western Part of the Upper Peninsula Michigan* and the highly unanticipated data taken during years of field research measurement activities.

ELF surely affected Earth's typical magnetism within our research areas, hence, our data in that geographic region changed as well. Too, it is easily proven that Earth's magnetism affects not only humans but also animals and vegetation. This must include carbon-based sasquatches as well. More on this later.

The Next Step in the Saga of Sasquatch and Me

Voila! Now we know what we didn't know then. Now we understand what was so befuddling at the time. Now we can differentiate the startling data we gathered between 1994-1998 from the radical changes measured in 1999. Even today, I feel conflicted at times as to whether I should thank or curse ELF for the many lessons it brought into our lives.

Regardless of the above, data gathered between the years 1994-1999 was valid then and remains valid today. Once Project ELF was permanently closed in 2004, the remarkable results we had obtained early on were no longer measurable in the same locations or in the same way. That is, while early results were reproducible over a period of years, the same results no longer held true commencing with late 1999. Let's not be depressed, though, for we now have even more fascinating data with which to contend.

That leads to this: Do not believe for a moment that obtaining the results we did in 1999 and the many years since meant an ending to our research. To the contrary. Once we determined ELF as the cause for the rapid and unexpected changing data in 1999, the next phase of our research began in earnest. After all, it would be incredibly difficult, if not outright naïve, to forget even momentarily our first odoriferous introduction to the phenomenon of sasquatch. Right? Well, this is right to me, and as I'm writing this; right is going to write!

Ultimately, and much to our benefit, we sorted through a large amount of data and confusion during the winter of 1999-2000.

Newfound clarity helped me develop a possible new premise: Our experience with the sasquatch in September of 1999 related to changing Earth magnetism that the ELF Project had created.

So, buckle in tight, grab a snack, snooze, or a brew, and prepare for even more wonders, because surely, they are on the way.

Chapter 2

[3] Federation of American Scientists (FAS). "Extremely Low Frequency Communications Program." Accessed February 17, 2018. http://www.fas.org/nuke/guide/usa/c3i/elf.htm. Site maintained by Steven Aftergood; originally created by John Pike; las updated 26 February 2004.

[4] Stromberg, Joseph. "Why the US Navy Once Wanted to Turn Wisconsin into the World's Largest Antenna." *Vox*, Vox, 10 Apr. 2015. Accessed 17 Feb 2018 www.vox.com/2015/4/10/8381983/project-sanguine.

[5] Koontz, Ronald L., Captain. "ELF Communications Systems." PDF. Accessed February 17, 2018. http://www.navy-radio.com/commsta/elf/elf-463-22A.pdf.

[6] "U.S. Census Bureau QuickFacts: Marquette County, Michigan." Census Bureau QuickFacts. Accessed January 03, 2019. https://www.census.gov/quickfacts/fact/table/marquettecountymichigan/PST045217.

[7] Olson, Daniel. "Clam Lake, WI ELF Transmitter." Military History of the Upper Great Lakes. October 11, 2015. Accessed February 27, 2018. http://ss.sites.mtu.edu/mhugl/2015/10/11/clam-lake-wi-elf-transmitter/.

[8] Villeneuve, Bradley. "ELF Station Republic, MI." Military History of the Upper Great Lakes. October 10, 2015. Accessed February 27, 2018. http://ss.sites.mtu.edu/mhugl/?s=ELF%2BStation%2BRepublic%2C%2BMI.

[9] **Note:** This file is a work of a sailor or employee of the U.S. Navy, taken or made as part of that person's official duties. As a work of the U.S. federal

government, the image is in the public domain in the United States. This file has been identified as being free of known restrictions under copyright law, including all related and neighboring rights.

[10] **Note:** This file is a work of a sailor or employee of the U.S. Navy, taken or made as part of that person's official duties. As a work of the U.S. federal government, the image is in the public domain in the United States. This file has been identified as being free of known restrictions under copyright law, including all related and neighboring rights.

[11] Wordnik. "Definition of Dipole." Wordnik.com. Accessed March 23, 2018. https://www.wordnik.com/words/dipole.

[12] US Navy. "Striving For A Safer World Since 1945." *Federation of American Scientists*, The United States Navy: Navy Fact File, 28 June 2001. Accessed February 17, 2018 https://fas.org/nuke/guide/usa/c3i/fs_clam_lake_elf2003.pdf, page 2. [].

[13] Merrill, Ronald T. *Our Magnetic Earth: The Science of Geomagnetism*. Chicago: University of Chicago Press, 2012, page 6.

[14] Ibid., page 2.

[15] Ibid., page 2.

[16] This diagram used according to the Creative Commons CC0 1.0 Universal Public Domain Dedication.

[17] Villeneuve, Bradley. "ELF Station Republic, MI." Military History of the Upper Great Lakes. October 10, 2015. Accessed February 27, 2018. http://ss.sites.mtu.edu/mhugl/2015/10/10/elf-sta-republic-mi/.

[18] Altgelt, Carlos A. "The World's Largest 'Radio' Station." PDF. Accessed 17 February 2018. http://www.hep.wisc.edu/~prepost/ELF.pdf [], page 7.

[19] "Northern Highland." *Wikipedia*, Wikimedia Foundation, 12 Apr. 2018. Accessed December 21, 2018, en.wikipedia.org/wiki/Northern Highland.

Chapter 3

Foolish Decisions, Errors, and High Hopes — 2000-2010

"I have not failed, I've just found 10,000 ways that won't work."

— Thomas A. Edison

Many people fear failure of any sort. Basically, it seems humans tend to not want to lose or fail at any endeavor. Don't misperceive where this is going, for I too dislike failure. Fortunately for me, I'm at a stage in my life where I view failure much differently than I did as a youthful, competitive, most likely conceited young snot. I now tend to view failure the way Thomas Edison's above-listed quote suggests. Of course, even though the quote clearly proposes a rational way to perceive failure as a teaching moment and an opportunity to move from decisions and actions that don't work to those that do, I wonder at times if Edison ever felt remorse for speaking those words. Hmm! He apparently upheld lofty standards.

Literature review regarding the sasquatch phenomenon uncovered an adage within the community of sasquatch believers: "Seeing is believing." I strongly adhere to this premise, although I strive to not use it very often because the adage seems to have become a commonplace, if not dreary, cliché that could easily transform into being meaningless from insufferable overuse. Merely an opinion; speaking of opinions, check out the next section.

Perception, Belief, and Reality

I would be remiss not to address how the concepts of perception, belief, and reality appear to interweave to co-create our personal existence—including the way we interpret unusual or anomalous phenomena. These concepts are particularly valid with respect to my sasquatch research, which you will readily see as we move through the book.

The subject of this chapter points out that foolish decisions, errors, and high hopes are also important to discuss. I strongly suspect that perceptions, belief, and reality intermix to drive decisions made and acted upon, along with errors acknowledged and corrected—or not, as the case might be. To reiterate, high hopes relate to expectations, while expectations in turn, power decisions made and acted upon. And, yes, this unquestionably includes my sasquatch research.

Interpretation of an experience with a bipedal anomaly is rather crucial and tends to drive how you respond or react following such an experience. For instance, if an anomalous experience deeply shakes your current belief system, you likely react to it. On the other hand, if an anomalous experience does **not** trouble your current belief system, you will likely respond rather than react. Bottom line: I suggest that long-held belief systems drive interpretations of and responses to anomalous experiences or, at the very least, greatly influences them.

As an example, I recall in the early 90s being involved in a conversation regarding a particular book. One of our friends—a brilliant, educated, successful, accomplished lawyer, and a player in a high-powered civil service position—asked the following question during our discussion, "Will reading *that* book challenge my beliefs?" I said, "Most likely!" She responded, "Then I don't want to read it." I was deeply struck at the time with the realization that many people I've known and interacted with over the years fear something similar. That is, it is not uncommon to not appreciate, not like, or certainly not want to have long-held beliefs challenged—for any reason.

I do not condemn this fear; neither do I support it. To the contrary, I believe if, during our daily life activities, we do not avail ourselves of opportunities to explore ideas that challenge our beliefs or stretch our thought processes away from a typical comfort zone into a previously unknown one, we are doing ourselves a bit of a disservice. Perhaps to the point of slamming the brakes on what many perceive as ongoing growth.

To wit, when I engage in behavior or thoughts that follow only deeply imbued traditional beliefs, it seems that a bit of temporary stagnation besets me and, stagnation for me creates so-called mind ruts. In turn, these mind ruts, regardless of their cause, stagnate my creative thought, and prevent me from moving outside a limited comfort zone, thereby holding me captive deep inside a box of imbued traditional beliefs. Although a well-established mind rut governed by only traditional beliefs may feel safe, warm, or fuzzy, it actually acts as a prison guard holding you back from exploring different or others' beliefs; beliefs that lie far outside those you typically label as safe. Stepping outside long held, deeply imbued beliefs, on the other hand, can serve to shake brain neurons into new and dynamic versions of belief. At that point, bye-bye long held traditional beliefs and hello new and stimulating thoughts and beliefs. Beliefs that open a door to a whole other you—or, perhaps, a whole other reality!

To that end, I suggest that my first encounter with a sasquatch created a rather worrisome, albeit thrilling challenge to my traditional belief system. I could have reacted in diverse ways, although two stand out as dominant—denial or acceptance. Obviously, I worked through the denial part and moved into acceptance or I wouldn't be tapping my keyboard while sharing this story. Even so, I assuredly had moments of self-doubt and self-distrust. While battling temporary outright disbelief, I moved rather awkwardly through many emotions in addition to the myriad thoughts that pelted me at the time of the encounter and for the next several months.

My life was momentarily turned upside down following my initial contact with a sasquatch because my traditional **beliefs** were challenged, and my concept of **reality** underwent a rather fierce adjustment period. What's the point of all this? Stated simply, the **perceptions** and **beliefs** I held at the time my sense of reality was challenged caused me to jump outside my traditional mind ruts into new, unknown thought.

Allowing new, unknown thought to reign—at least temporarily—created a scenario wherein new, unknown thought drove my **interpretation** of my changing **reality**. Finally, my interpretation of continued anomalous experiences served to direct my decisions about what to do next. And yes, I was definitely unsure what would happen once I started down a new, unknown trail driven by different thoughts and experiences. This was singularly scary and stimulating. I hoped I wasn't stumbling into a deeply, unknown rabbit hole!

Once things settled down a bit, I made a few errors, engaged in periodic poor judgment, and screwed up from time to time as I bumbled and stumbled my way into intentional research of the sasquatch phenomenon. The next two sections are divided into timeframes; first, I address the winter of 1999-2000, and then the research years from 2000-2004. Some of my errors and foolish decisions are mentioned there. However, I'm pleased to say that neither errors nor foolish decisions prevented me from steadfast forward movement into a novel and exciting research for the ever-elusive sasquatch.

Winter of 1999-2000

I read and tried to assimilate all that was humanly possible about *bigfoot* throughout the winter of 1999-2000. I was quite enthusiastic, and dutifully sought to unravel any, if not all, mysteries surrounding our first sasquatch encounter of 1999. I bought and read book after book after book about sasquatches. As well, I downloaded and printed hundreds of pages from online sites and blogs about the bipedal anomaly we had smelled, heard, and seen. Basically, I informed myself following early struggles with perception, belief, and reality.

At that time, I was adamant that I would discuss this experience only with those who were with me or with whom I felt quite close. So, I did not speak of or share this experience with anyone outside of my tightly held circle of friends. As you might expect in this situation, not addressing this issue outside of our enclosed circle was potentially less than wise. Why? Because no one within our circle knew a damned thing about sasquatches! This meant, of course, that whatever conclusions we drew were bound to be somewhat narrow and lacking in that always present wisdom maker—experience. I did not, though, recognize this at the time. I was so paranoid that folks would treat me as an outcast that I did not detect the wisdom in speaking with someone who had experience with this anomaly. Perhaps periodically assessing hindsight is not a bad approach to realigning one's sense of shifting realities.

It wasn't until the end of my late Fall 2017 expedition (which lasted nearly six weeks) that I recognized the need to place almost two decades of sasquatch research into the public realm. The shocking, nearly unbelievable results of that expedition appear later in the book. As with all stories, chronology is important for, without chronology, drawing conclusions seemed befuddling or maybe pointless. Anyway, although I struggled greatly with the idea of placing my sasquatch research into the public realm, in the end, I chose to do so, and here it is. My friend Julie, who you will meet a bit later, first prompted me to consider writing a book, and she has stood by my side throughout the entire process!

Finally, the summer of 2000 approached, and I and my friends spent hours discussing best possible approaches on how to continue my heart/head research while increasing efforts to search for a bipedal anomaly. Nothing wrong with that, or so I thought at the time. Sadly, a problem arose with this approach as I wrestled with the question of how many and who would be included in the new category of sasquatch research.

I did not wish to include new people, mostly out of fear of severe ridicule or worse, so it became a bit challenging to work the heart/head research *separate* from the bipedal anomaly research when subjects for the heart/head research were not included in the bipedal anomaly research.

Whew! In the end, engaging in ongoing research in this manner did indeed serve as one of the foolish decisions I made. This will become clearer as we look at the years 2000-2004.

2000-2004

This period turned out to be a gung-ho attempt to interact with a bipedal anomaly and find a way to prove two important factors: (1) sasquatches exist, and (2) they must be something we could reasonably determine as less than human yet more than animal. No point jumping into the search for a sasquatch with low or no expectations, eh? In addition, I expanded my research into the heart/head theme I'd been following for five years. Talk about a heaped plate! Imagine the indigestion I experienced gobbling all this down over a limited research period. Talk about *heart* burn and *head* ache!

Our pattern for 2000-2004 was to travel to a research area for a week in July, a week or 10 days in August, and at least a week in September. We kept this routine through most of 2004. Even though the ELF Project discussed in Chapter 2 altered our heart/head aggregate data a bit, we continued to add support to our earlier theories of how changing Earth magnetism affects humans. This became particularly obvious with respect to the autonomic nervous system, particularly with respect to the heart/head connection. In fact, we quite successfully pursued this avenue as our collective data persistently supported earlier premises. We simply needed to find adequate locations where changing Earth magnetism occurred.

I also dedicated time each trip to ongoing research into our bipedal anomaly phenomenon. For instance, I began a different activity in 2000 after discovering the idea of track casting prints. I had read about track casting, bought plaster of paris, and spent time at home practicing the art of track casting by following instructions downloaded from an online site. Gloomily, my home practice did not effectively assure successful track casting in the field. Even so, upon finding bipedal anomaly footprints in the wilderness area of our research, I excitedly casted them and ended up with—you got it—a few essentially useless casts. Apparently, I was still learning how to implement forensics techniques.

I continued my pursuit of literature review over the winter of 2000-2001. More books, more online sites, more sorting out and organizing what I had read, and, finally, striving to figure out how to put what I learned to respectable use in the field during the following summer and fall.

I also spent hours and hours creating charts, tables, diagrams, and analyses summaries for the heart/head data. This data pumped me up because it continued to lend insight into, and proof for, our heart/head research premises. All-in-all, another exciting and education-packed winter.

In July and August of 2001, we once again drove north from our mid-Michigan home to pre-selected research areas. We all remember the tragedy that happened in September of 2001; no trip north that fall. Also, due to my ongoing research with the heart/head connection interfacing with changing Earth magnetism, I had scheduled a training jaunt out to northern California. I wanted to learn more about the head/heart connection, and intended to take a training workshop from an institute that had been researching the head/heart connection since 1990—the HeartMath Institute.[20] The training was scheduled for mid-September 2001. Due to the terrorist activity on Sept. 11, 2001, the HeartMath training was postponed until November 2001. You may recall that not a single plane graced the unblemished skies across America right after that heart-breaking catastrophe.

I had read everything HeartMath Institute had published or made publicly available online. I had several phone conversations with one of their HeartMath coaches while completing training over the phone. In short, I was both excited about, and impressed with, the HeartMath material and their scientific research about the head/heart connection, the autonomic nervous system, and an exercise they developed to aid the process of so-called head/heart coherence. Again, I mention coherence, and will continue to do so throughout the book. I assure you, as we approach the end of the book, the idea of coherence will become quite clear to you. So too, will the reason I'm so fond of using coherence during my field research excursions.

HeartMath research was light-years ahead of other researchers at that time and I was thrilled to have taken part in their training program. To this day, I remain faithful to the exercises I learned during my training. I also stay abreast of their continued and expanded research programs. Though it might seem a bit odd for me to address HeartMath as part of this story, believe it or not, what they taught me plays a vital and ongoing role in this sasquatch and me story. Go figure! Sometimes the most unlikely tools land in our laps, thereby aiding in producing phenomenal results. Only, though, if we remain open to new and dynamic beliefs.

Before I jump into the next section, it seems proper to mention here that Brushtracker and I caught a glimpse of a bipedal anomaly in the fall of 2004—our **second** glimpse of a sasquatch. We had been hiking in a wooded, hilly area, noticed footprints, and decided to follow them for a bit. About 20 minutes later, we were in a relatively short ravine, maybe 200 yards long, when we heard something snapping twigs while moving slowly through the woods at the opposite end from us. We stopped to listen, looked around, yet saw nothing; nor did we smell anything.

As we began moving again, we thought we saw movement near the top of the far end of the ravine. Not able to see things clearly, we kept slowly moving in that direction. As we moved closer to the point where we would need to climb to get out of the ravine, we heard the same noises we had heard earlier. Again, we stopped to listen and look around; once more, we saw nothing. Clearly, whatever was causing these noises was not being stealthy; ergo, we decided that it meant for us to know it was there, whatever **it** might be!

Just as we began our climb out of the ravine, we heard yet another unusual noise. The only way to describe what we heard then is to label it as a quiet chuckle or chortle. It seemed to originate directly above where we were standing, so we held steady in our position, turning only our heads and eyes while striving to catch a glimpse of the chuckler. After about five minutes of this, and without success, we slowly continued our climb toward the top of the ravine. Made me wonder if that *something* was chuckling at the way we floundered through the forest while looking for it! Hmm! Maybe sasquatches have a sense of humor?

Keeping our eyes directly ahead and upwards we walked with slow, intentional steps, not speaking as we climbed. In short order, we caught movement just to our right and at the top edge of the ravine.

Wait! We pointed simultaneously at the object we saw moving. It was clearly another bipedal anomaly, although we saw much less of this one than the sasquatch we had seen in 1999. We recognized the back, arms, and shoulders of this one as it moved away from us. It was not in a hurry; however, the brush was quite thick, so we were unable to gain an open or clear view of it. As it moved deeper into the surrounding woods, we heard it make the chuckle/chortle sound again—and then nothing. No twigs snapping; no brush sounds as it passed through; no further vocal sounds; nothing. It was as if it had stepped into another reality; one that didn't include us. In other words, we were amazed that something so large could move so stealthily through the dense thickness of the wilderness. Wow! Compared to it, Brushtracker and I must sound like steam locomotives barreling through the forest! The only thing missing was an obnoxious steam whistle.

Thus, our second anomalous experience ended. Were we excited? Indeed! Were we pleased that each of us had similar observations? Yes! Are we able to offer definitive, provable facts about the experience? Nope! That is, other than to say that this creature had seriously broad shoulders, and what looked to be about 3½ inch long brown hair on its back and arms. We were unable to gauge its height, although it was obviously quite tall. And again, it was **very** strong. This was our **second** view of a sasquatch. This, though, offers a peek into the upcoming section because both our bipedal anomaly experiences surely fall under the heading of an a*nomalous experience*—possibly intertwined with something known as *cognitive dissonance.*

Anomalous Experience and Cognitive Dissonance

The online *Merriam-Webster Dictionary* puts forth that an anomalous experience is something "inconsistent with or deviating from what is usual, normal, or expected; irregular or unusual."[21] Anomalous experience experts, Thomas Rabeyron and Tianna Loose from the University of Nantes in Nantes, France wrote in a 2015 online article, "Anomalous experiences occur when a person is confronted with an experience that does not fit with a usual conception of reality, an 'extraordinary' interaction with the environment that can produce both a positive and a negative impact on mental health."[22] Uh, oh! Don't care for the sound of that!

Next, let's look at a common definition for cognitive dissonance. According to Wikipedia, "In the field of psychology, cognitive dissonance occurs when confronted with new information that contradicts … beliefs, ideals, and values.[23] I seriously doubt if a strong argument can be made suggesting that an experience with a bipedal anomaly does **not** qualify as an extraordinary interaction with the environment which, in turn, might trigger a temporary bout with information that contradicts traditional beliefs, ideals, and values.

I'm quite sure these two concepts played out in me following my experience with a bipedal anomaly in 1999 and again in 2004. I'll not comment here on how this may have played out for Brushtracker. We didn't discuss this aspect of our experiences at those times. Rather, feeling somewhat unsettled internally, I began research on how folks might reflect upon an anomalous experience after it was over.

I pursued this theme from a perception and belief perspective that included the human biological system. I discovered that HeartMath's lead researcher, Rollin McCraty, wrote that, "throughout the body, information is encoded in *waveforms of energy* as patterns of physiological activity"[24] (author's emphasis).

I should point out that McCraty has guided HeartMath's research program over the years to a point where he has become world-renowned for his meticulous scientific work, publications, and books about the physical and psychophysiological head/heart connection. That said, McCraty also supports the premise that, "neural, chemical, electromagnetic, and oscillatory pressure wave patterns are among those used to encode and communicate biologically relevant information."[25] This is another reference to *wave energies*, as well as ideas relating to the human nervous system, chemical interactions, and electromagnetism. I shall continue to build on these ideas throughout the book because they serve as a foundation for later premises. Furthermore, the ideas implying chemical interactions offers support for a hypothesis known as Photosynthetic Piezoelectric Induced Transparency (PPIT). This theory, discussed in Chapter 16, was put forth by a longtime sasquatch researcher, Pearl J. Prihoda.

McCraty also made known that, "... the heart plays a central role in the generation and transmission of system-wide information essential to the body's function as a coherent whole."[26] Let's focus for a moment on the earlier stated idea that our physical bodies undergo neural, chemical, electromagnetic, and oscillatory pressure wave patterns. This thought bespeaks a clear and direct correlation between the human body and external magnetic influences. This is especially true when combined with HeartMath's demonstrative research showing how measurable magnetic qualities of the physical heart *extend outward* from the physical body at least 10 feet. By the way, this premise has been scientifically confirmed over many years by many researchers in diverse studies and settings.

Imagine for a moment that you are standing or sitting in a physical location where Earth's measurable magnetic energy is changing. Momentarily set aside the question of what causes Earth's magnetism to change, and focus instead on how the physical heart's measurable magnetic qualities extending outward from the body might respond to correlative changes in the earth's magnetic field. Stated simply: As the earth's magnetism alters, there **must** be a corresponding reaction within the physical heart's magnetic field. This is precisely what years of our heart/head research data verifies. I believe this also applies to sasquatches because they are carbon-based beings similar biologically to humans. More on this later in the book.

Considering that McCraty's premises were scientifically proven many times over, I boldly suggest it follows that *changing Earth magnetism creates a corresponding change within the physical body, including the neural, chemical, electromagnetic, and oscillatory pressure wave patterns of the physical body*. Or so it seems to me. Furthermore, HeartMath's thoughts on coherence present an opportunity to use coherence techniques to offset or improve confusion occurring following an anomalous experience, or, perhaps, keeping one from dipping one's toes into cognitive dissonance. Again, more on this as we move through the book.

To summarize this section, I believe each of us engages in, understands, and reacts to so-called anomalous experiences differently, due in some measure to the way we **perceive** the biological tweaks occurring within us during an anomalous experience, and perhaps, immediately following such an experience as well. I further suspect that our **beliefs** at the time of an anomalous experience foretell in part the way we initially **interpret** an anomalous experience. It seems reasonable, then, to recognize that our **perceptions**, blending with beliefs, helps define our **reality**. So, here we are; back to considering the concepts of perception, belief, and reality, gently intermingled with the idea of interpretation.

Extending McCraty's established premises, I also put forth for consideration that each experience we have—regardless of what it might be, or the cause—triggers a corresponding internal physical response, with variant outcomes due to our perception of each experience blending with the beliefs we hold at the time of the experience. The HeartMath Research Center suggests that its research since 1990 supports an *actual* physical/psychophysiological manifestation within the human body during each experience in which we engage. In other words, everything we do—our thoughts, our emotions, our actions, our behaviors, our beliefs, etc.—produce corresponding physical and psychophysiological results within the human body. Rather startling, wouldn't you say? And yes, this is pertinent to the story of sasquatch and me, which includes addressing anomalous experiences and cognitive dissonance.

In a way, this important idea leads us full circle back to our first research interests. That is, the heart/head connection, and what might happen when we intermingle changing Earth magnetics that cause a corresponding physical experience with guidelines attached to what HeartMath calls coherence. A bit abstract? Yes, for now; later, though, this idea moves from the abstract into a readily understandable format. Patience, please.

It seems obvious from HeartMath's research, as well as our field data, that a clear and measurable effect upon the human body occurs when it is introduced to a location undergoing changing Earth magnetics. This premise has been checked, measured, well-documented, and reproduced time and again throughout our years of heart/head research. Please bear these thoughts in mind for, shortly, these thoughts lead to a discussion on the value of learning more about HeartMath's research on coherence and how helpful intentional coherence is to the entire human system, not to mention offering a way to successfully interact with sasquatches.

I suspect a few reading this material might think I'm taking a rather circuitous trajectory just to arrive at certain conclusions. Well, I am; and there is a reason for doing it this way. Regrettably, the reason will likely not be clear until a bit later in the book. Therefore, I continue to ask for your patience until we reach the point of pulling this information together into a clear, concise conclusion. Pending that, a couple of summarizing questions:

1. **Question:** Is it possible that sasquatches have a similar, measurable, physical heart magnetic field—as do humans?
    a. **Answer:** Yes!
2. **Question:** If so, would they not also experience interactions with changing Earth magnetics?
    a. **Answer:** Yes!

2005-2007

The timeframe of 2005-2006 saw us conducting research in July and September each year. We had many experiences during this time involving twigs snapping, somewhat noisy movement through the wilderness, and even periodic vocalization sounds. We heard whistles of different pitches, various grunting sounds, as well as a variant on whistles. I believe that two of the variant pitch sounds were chuckles or chortles. We were also introduced to something known in the sasquatch arena as wood knocking. Fascinating! Throughout this time, we continued to discover and follow footprints; way too many to count.

We took scads of photos over the years, including a few that fall squarely into the round hole of so-called *blob* photos. While in the throes of an anomalous experience, one might easily perceive blob photos to show a sasquatch, yet they show nothing that would contribute to definitive identification of a bipedal anomaly. Guess it can't hurt to pump up the adrenalin from time to time, though, so we certainly will not cease slapping fingers on the camera's snap button. One outcome of note following years of bipedal anomaly research; we became quite adept at finding, categorizing, and following footprints. This skill certainly contributed to later successful experiences with sasquatches.

Continuing, the year 2007 found me trekking into the wilderness alone at times; a rather radical change from my typical patterns. Although I followed proper precautions for safety, my wife and many of my friends were a tad distraught with me for going out alone. Some of them persist in their unrequited dissent to this day. Even so, my nearly six-week expedition in late fall of 2017 confirmed to all involved that when I'm alone, I have profound encounters with bipedal anomalies. More on that in chapters 12 and 13. Shocking stuff, I say!

On Becoming A Woodsman

I learned my woods craft from an uncle, starting at age 15. He was an avid hunter, a superb woodsman, and an excellent tracker. He taught me the art of tracking, carefully listening to all the sounds of a wilderness, and how to develop and constantly enhance strong wilderness observation skills using all my senses. He taught me to truly see, smell, hear, and sense all that a wilderness offers a visitor. Additionally, he taught me the art of patience while wandering through a wilderness. I remain indebted to him for his willingness to teach me; without his lessons, my field research would not be what it is today. Thanks much Uncle Bob!

Bottom line: I do not consider the decision to spend time alone in the wilderness one of the errors or foolish decisions referenced by this chapter's title. To the contrary, going out alone has turned out to be one of the wisest choices I've made in this pursuit of gleaning knowledge about some of the habits of sasquatches.

This then, returns me to the idea of coherence and how coherence aids one's decision making skills as well as being beneficial during and after a bipedal anomalous experience. In brief, coherence is, first and foremost, a physical experience; it leads, though, to rather thoughtful happenstances through heightened senses, mental clarity, and emotional stability. It is indeed a powerful tool!

Coherence

I was intrigued when I first became aware of HeartMath's research on coherence. Then, I dove into the deep waters of its scientific research and swam with currents of analysis to see if I could fathom the meaning of coherence as it might relate to my research. This, because the foundation of HeartMath's claim assured that implementing this thing called coherence creates physical control of electrical signals flowing between the heart and head. Say what?

Are you beginning to understand why I say that coherence is powerful? I wondered if adding coherence techniques to my research routines might augment my investigation into how changing Earth magnetics affects humans. This turned out to be a respectable and plausible current of research for me to swim in—and at times in which to float gently along for a bit when the material or experiences became temporarily overwhelming.

HeartMath's McCraty addressed heart coherence in a 2016 released book entitled: *Heart Coherence: Connecting with the Intuitive Guidance of the Heart.* He wrote about HeartMath's discovery of a physiological state they named heart coherence, claiming it "supports a balanced partnership in the interactions between our heart, mind, emotions, and physiology."[27]

Stated simply, heart coherence involves a physical interaction between the physical heart and brain via the autonomic nervous system. Ultimately, coherence induces a verifiably balanced physical response between the heart, brain, and autonomic nervous system. Basically, heart coherence purports we can take literal control over this aspect of our physical experience. Rather mind-blowing to imagine such a thing, eh? Yet, I have proven it works quite well—even during sasquatch research!

You might still wonder how this seeming nonsense relates to a search for sasquatches. Again, a request for patience, for I assure you, **heart coherence now controls my personal encounters with bipedal anomalies**. So much so, that practicing heart coherence lengthened each added contact with a bipedal anomaly. That is, once I engaged in a regular practice of heart coherence while in the wilderness, my physical encounters with sasquatches now last minutes, much longer than the mere seconds of earlier such meetings.

As we progress more deeply into the realms of magnetism, geomagnetism, space weather, and heart coherence, this idea becomes much clearer and more pronounced. For example, see Chapter 4 for a brief discussion on magnetism, geomagnetism, and the efficacy of heart coherence.

2008-2009

The 2008-2009 timeframe found us traipsing through the wilderness essentially chasing our own tails. No, I'm not proud to admit this; however, by the end of 2009, I recognized that it was time for me to solidify or come to grips with my intentions, hopes, and desires such that I establish and adhere to a consistent approach for my ongoing sasquatch research. This led me to rethink, reevaluate all data to date, reorganize, and then engage in what I hoped would be a well thought out continued exploration. The 2008-2009 period produced nothing other than scads of footprints, assorted noises, and a few photographs—with one notable exception.

In late summer of 2009, Brushtracker and I had yet another bipedal anomaly experience. That experience caught us off guard; plus, it had an unusual ending. The same group of five that took part in the 1999 odoriferous encounter rejoined for a wilderness excursion in mid-summer of 2009. We came across a flush blackberry patch on a high hill in Iron County, Mich. It was rife with fresh, delicious blackberries; we stopped to pick a couple capfuls to take back to camp for desert. Yum!

After picking several capfuls, we heard something at the west end of the hill we were on. Brushtracker and I walked that way until we were able to peer down a steep hill; Maggie, BoB, and Dragonfly continued to pick berries and chat about our day's experiences. Brushtracker and I were on a ledge of the hill where the hill was much steeper and brushier. As Brushtracker and I peered downward, we saw what looked like a bipedal anomaly. Sure enough, it was!

Standing smack next to a tree with its left shoulder behind the tree as it watched us, we could see its right shoulder, right arm, and the right side of its face. We could not see any other part of the body. It was standing approximately 35 yards downhill from us and it was looking directly up the hill at us. It stood perfectly still, and we felt its right eye locked onto the two of us.

An odd sensation, to say the least. This time, however, we were not startled; rather, we quietly questioned how we might adjust our position to see more of it without it running away. Of course, my camera was back by the berry patch! Will I never learn? As we watched it and it watched us in return, we saw a little movement.

Also, the sun was out and shining brightly toward us, which put sunrays at its back. For lack of better terminology, we noticed a shimmering around the tree where the bipedal anomaly stood. The shimmering was like that which can be seen when observing heat energy rising above a hot asphalt road.

The shimmering intensified for a moment; then, when it cleared away, we saw nothing. At this point, we looked at each other and shook our heads. I said to Brushtracker, "I know I saw it; he replied, "I saw it too." After a quick discussion, we decided to trek down the steep hill to the tree where we saw the bipedal anomaly.

I was convinced we would see footprints at the base of the tree. We climbed carefully down the steep hill, using trees and branches to steady our descent. We reached the tree, looked at the base of the tree, and … good grief! … not a print to be seen. Nothing. No sign whatsoever that anything had been standing there!

Okay. Now we were a bit startled because that made no sense. We decided it would be easier to keep going downhill than to climb back up from that precarious position, so we did. At the base of the hill we stepped into a ravine and chose to follow the ravine toward an easily walkable trail that would lead us back uphill. As we moved across the ravine, we discovered a huge pile of dung. When I say huge, I mean way larger than human dung and way more quantity than humans produce. Plus, it was seriously fresh, almost as if the bipedal anomaly we had seen took a dump before it started up the hill toward us. My guess is that it was headed for the berry patch we had inadvertently invaded. Oops!

By this time in our research for bipedal anomalies, I carried latex gloves, plastic evidence bags, and tweezers with me for the times that collecting evidence worth taking home for analysis availed themselves. I gathered a bit of the dung to take back with us. We looked around and saw a bunch of footprints, although none were castable. My camera was topside on the hill, so no photos at that time. Apparently, I was still prone to foolish errors.

We walked back uphill, shared our experience with Maggie, Dragonfly, and BoB, and surprisingly, they believed us this time. Dragonfly had noticed hair among the berry bushes, so I gloved up again, extracted it, and placed it in a separate evidence bag, hoping to get it analyzed upon our return to civilization. Although the dung ended up being thrown out, I still have the hair after all these years. It could very well be bear hair because that area certainly has bears tromping through the wilderness. On the other hand, we really don't know.

Back to camp for the night to enjoy the luscious berries. The following morning, Brushtracker and I headed back out to the same site. It was raining, so the others chose to drive to town. Brushtracker and I arrived back at the spot where we had collected the dung and noted a very odd thing; something had *ravaged* the bottom of the ravine in what appeared to be an intense rage. At least a 70-yard long area that was 20 feet wide had been violently stripped and mauled. Small trees had been twisted off midway up from the ground and tossed aside, brush was madly destroyed, and what looked like huge prints had pummeled the ground, although none were track castable.

Interestingly, the photos I took of the destroyed area are missing at the time of this writing (February 2018). I've looked everywhere for them and am unable to find them. I had hoped to put one or two photos of that ravaged area in the book. Bummer! One thing for certain; Brushtracker and I are in full agreement we would NOT want to be anywhere near whatever it was that caused that type of destruction while it was in an uncontrolled rage. Geesh! Unless, of course, we could be invisible. Shudder … shiver!

Moreover, we do not know why something wanted to demolish this area with such intense rage. I suppose the culprit who did it might have been the bipedal anomaly we had seen the day before. Perhaps it resented our presence in the berry patch. Maybe it was angry because I gathered dung. Or, something else could have triggered such rage after we left the area. Once again, we'll never know for sure what caused such intense rage. Regardless, seeing the amount and force of the destruction certainly put a thoughtful twist to our ongoing search for bipedal anomalies. This then, was Brushtracker's and my **third** visual encounter with a bipedal anomaly. To remind you, we saw one in 1999, one in 2004, and again in 2009.

2010

The year 2010 opened me to many realizations, including the reality of mortality. In June 2010, I packed my tent and camp gear into my Jeep and headed west for a three-week excursion into the Rocky Mountains. I found a small, rustic national campground that had few spaces, and the spaces were far apart, separated by whistling pines. I was at the end of the park where I camped on the powerful Teton River in Montana. During the week, I was the only camper and only three or four campers appeared on the weekends, so I was mostly alone.

I set out on foot to discover the wonders of the western mountain wilderness. Such amazing beauty; such breathtaking scenery; such unequivocal and heady realizations of self alone within a vast wilderness. For example, one day's hike up a mountain found me looking at a handwritten sign placed by a forest ranger. It read, "Carcass ahead near the lake. Beware of Grizzlies." No worries. As the sign was obviously a couple of days old, I brazenly chose to pursue the path ahead, thinking the carcass was long gone by now. After about 20 yards along the trail, I heard what sounded like something being dragged somewhere. It was just around a curve in the path—perhaps 70 yards ahead.

At this point, I shamelessly turned around and rather quickly—albeit quietly—walked in the opposite direction. Not suggesting I was scared; rather, that my heart was pounding fiercely and loudly enough for creatures in Wyoming to hear it. Whatever was around the curve must have thought loggers hit the forest. Oh! I said a bit ago I was alone; seems as if I guessed wrong!

One early morning before leaving camp, I was drinking a final cup of coffee at the campfire. An elderly gentleman dressed in well-used outdoor hiking gear walked by, stopped, and said, "Good morning." I responded, "It is, indeed, sir," and asked him if he'd like a cup of coffee, to which he responded "Sure." I poured him a cup and we sat by the final burning coals in the firepit.

He engaged me in conversation by asking many questions about what I was doing, why I was alone, and so forth. I told him about my love for the wilderness; however, I did not mention my search for a bipedal anomaly. After chatting until our coffee was finished, he said he was quite familiar with the area and he rarely spoke with strangers. He then mentioned to me he could direct me to a location where I would see an eagle's nest with young ones in it. Cool!

He gave me explicit directions and then left my campsite, fading quietly into the surrounding wilderness; and I do mean quietly. Nearly spookily so! Clearly, the gentleman was not a stranger to the ways of the forest.

I got into my Jeep and drove to the spot he suggested. I stepped into the wilderness, followed his directions and, within 15 minutes of hiking, came across bipedal anomaly footprints that were positively **huge**. Although I measured them, I won't write here what the measurements were because I find it difficult to believe them myself. Seriously!

I followed the tracks for a bit because they went in the same direction that he suggested I go. About five minutes later, I came across a small pond and watched two beavers swimming synchronously back and forth. Such symmetry! Such impeccable timing with their turns and dips. I felt I was watching a wilderness love story unfold. My gosh, they would touch noses each time they reached the end of the small pond and were ready to turn for the trek back. I was, of course, amazed and felt great happiness fill my heart. To this day, I can see them in my mind's eye. Such a joy to behold!

I continued to follow the stranger's directions but did not find the eagle's nest. He had expressed to me how much he *hated* cameras and had asked me to not take photos if he shared with me the location of the nest. I honored his request, so I do not have photos of the footprints or the beavers. At least this time, not taking photos was intentional.

Basically, things went well during this trip, and being alone in camp, I had time to ponder what steps I needed to take with my homeland sasquatch research. It turns out this road trip was exactly what I needed. I was able to collect my thoughts and turn them into a plan for organized future action.

Once a new plan was put in place, my continuing search for sasquatches took on different hues of success. Chapter 4 begins discussion on the novel approach I chose to implement, and the rest of the book follows suit. I feel exhilaration just thinking about what is forthcoming. Hmm! A bit of egocentric adrenaline? Perhaps.

Oh, my! In mid-July 2010, I had a heart attack, hence, was not able to make it into the Michigan wilderness for an extended stay that year. That would have to wait until 2011. Even so, while recovering from the heart attack, I did manage to put in a week up north that fall. Maggie and I rented a cabin as I lacked the strength for rough camping Not much was done because I was unable to exert myself physically. Nonetheless, the week I spent there allowed me to hone plans for the 2011 expedition.

I wasn't aware at the time that the eventual unfoldment of the new plans I drew up meant I would leave the western Upper Peninsula of Michigan and move into an entirely different location into the Northern Highland geographic area of northern Wisconsin. That awareness came in 2011. The end of 2011 concluded with me gaining a bonanza of insights, as you shall discover in Chapter 4. Before we go there, though, it's time to make sense of this chapter's wanderings into diverse thoughts.

Weaving Nonsense into a Pattern of Understanding

To bring the potential disparity of this chapter to a close, I'll start with the subject of the chapter: foolish decisions, errors, and high hopes. I had made two decisions that readily tie for first place in the race for foolish decision-making. First, not seeking out and speaking with someone who had years of experience in the field of sasquatch research. Second, conducting two simultaneous research programs with so few bodies and too little time. Both decisions rise to the top as being foolish; both have since been corrected.

Although I ingested literally thousands of pages of sasquatch information in a relatively brief period, I lacked the time to create and follow a dedicated pattern of bipedal anomaly research; hence, that too was a bit foolish. I also learned that I had not allowed adequate practice time to develop proper track casting techniques. Even though this is not a major issue, it falls under the heading of being an error. One of the more foolish errors in which I engaged during the pilot stages of bipedal anomaly research was to not look *up* while in the forest. Silly me! I, of course, am clueless as to how much I may have missed in the early years due to this error; this error, too, has been corrected.

Interestingly to me, I discovered that shortly after the bipedal anomaly encounter in 1999, I jumped headlong into the arena of high hopes. Perhaps excessively so; and with serious high hopes holding reign, they acted as a driving force for many research decisions. Therefore, I suspect excessive high hopes may also align at times with foolish decisions or, perhaps, as catalysts for errors. I say this because I did not recognize my excessive high hopes for what they were at the time they occurred and was unable to at once rein them in to a manageable level. Again, I now manage my hopes, desires, and actions more wisely than before; I believe the rest of the book will verify this assessment.

Learning coherence aided me in an eventual sorting out of foolish from wise decisions. In turn, this helped govern changes in research patterns, as well as decisions for future efforts in the wilderness. I remain astonished at how coherence is truly an all-around pertinent and useful tool.

I believe a newly acquired understanding of anomalous experiences combined with an eventual blending of possible cognitive dissonance aided me in working through my first three bipedal anomalous experiences. Perhaps there are times when a simple awareness of various concepts serves to assuage future issues even before they arrive. Just a thought.

For me, a central part of this chapter relates to the role of magnetism in my ongoing research. I introduced it within a twofold setting: First, changing Earth magnetics affecting humans, and second, how the physical heart exudes a measurable magnetic field at least ten feet from the physical body. Tantalizing as this is alone, more is forthcoming because, in due course, magnetism—as well as geomagnetism and space weather—play serious roles in my research in ways I had not previously read or heard about while engaged in ongoing sasquatch literature reviews. Hopefully, you will find the magnetic aspect of my research as tantalizing as I do.

I believe the crux of this story about a search for an elusive sasquatch, and the few successes we've had, revolve around Earth's surface magnetism interfacing with geomagnetism during geomagnetic storms. Hence, I suggest that Earth's magnetism intertwined with geomagnetism blends with electromagnetic qualities of the human body in ways that allow for different and, in my case, successful interactions with sasquatches. This is a rather bold statement for sure; however, please continue reading to follow my thought process and the way events from 2011-2017 play out. Those years were intriguing and filled with amazing experiences.

In the end, I **recognized** my errors, **acknowledged** them, **accepted** responsibility for changing them, and **responded** by moving forward with a well thought out plan for continued field research. Following that, I hit the field, plunging into even more dynamic and thrilling anomalous experiences—minus the confusion of cognitive dissonance.

Chapter 3

[20] HeartMath® is a registered trademark of HeartMath Institute.

[21] "Anomalous." Merriam-Webster.com. Accessed February 23, 2018. https://www.merriam-webster.com/dictionary/anomalous.

[22] Rabeyron, Thomas, and Tianna Loose. "Anomalous Experiences, Trauma, and Symbolization Processes at the Frontiers between Psychoanalysis and Cognitive Neurosciences." *Frontiers in Psychology*, Frontiers in Psychology, 21 Dec. 2015, www.frontiersin.org/articles/10.3389/fpsyg.2015.01926/full.

[23] Wikipedia. "Cognitive Dissonance." *Wikipedia*, Wikimedia Foundation. Accessed February 22, 2018, en.wikipedia.org/wiki/Cognitive_dissonance.

[24] McCraty, Rollin, et al. The Coherent Heart: Heart-Brain Interactions, Psychophysiological Coherence, and the Emergence of System-Wide Order. Institute of HeartMath, 2006, page 5.

[25] Ibid., page 5.

[26] Ibid., pages 5-6.

[27] Childre, Doc Lew, Howard Martin, Deborah Rozman, and Rollin McCraty. *Heart Intelligence*. Cardiff, CA: Waterfront Press, 2016, 67.

Chapter 4

Magnetism Dances Blithely with Geomagnetism — August 2011

"... solar and geomagnetic activity can affect human health and behavior, including social behavior and unrest around the world."

— Annette Deyhle

This chapter delves into the saga of the 2011 expeditions, presenting early findings related to magnetism, geomagnetism, and space weather. It's time to shamelessly dip our toes into blossoming theories about these findings and sasquatches. Too, I flirt with an introduction to phenomena known in the sasquatch arena as *habituation* and *familiarization*. In the end, habituation and familiarization became a dominant factor in future research efforts following 2011, so please do keep this thought in mind.

The year 2010 saw the end of my changing Earth magnetics affecting humans research. Starting with 2011, I dedicated my time and efforts to expanding the bipedal anomaly research. I engaged in two expeditions in 2011, uncovering fascinating sasquatch information via measurement of various magnetic fields worthy of mention. First, though, I need to explain how we transitioned from the western Upper Peninsula of

Michigan a bit south to the deep wilderness of the Chequamegon-Nicolet National Forest, which is two actual forests—Chequamegon and Nicolet—combined in 1993 for purposes of forest management. Also, and for your information, the word *Chequamegon* is pronounced with a silent "q."

As mentioned in the prologue, obtaining the *Total Intensity Aeromagnetic Map of Western Part of the Upper Peninsula Michigan* was crucial to our research. Secondarily, obtaining many other maps greatly enhanced our ability to pursue our research with expedience and directed intent.

While Maggie and I were in the western Upper Peninsula of Michigan during the fall of 2010, I spent hours perusing our maps because I was unable to hike far while recovering from my heart attack. One of my U.S. Geological Survey (USGS) topographical (topo) maps included the Chequamegon-Nicolet National Forest located in the Northern Highland geographical area of northern Wisc. I immediately noticed quantitative topographic, flora, and magnetic similarities between Michigan and Wisconsin, which led to a more thorough examination of available maps.

After several hours of drawing comparisons between the areas, I decided to obtain additional northern Wisc. maps for review. I ordered a few maps and, over the winter of 2010-2011 invested hours diligently poring over those maps. By the spring of 2011 we decided to take a brief reconnaissance trip into that area to check it out.

Following that trip, I decided to take our equipment and begin exploration of the Chequamegon-Nicolet National Forest in the Northern Highland geographical area. Once again, with a focus on magnetism, a magnetic map for the area was obtained. As with the magnetic map for the western Upper Peninsula of Michigan, the Wisconsin magnetic map became paramount while defining potential research locations. As it did in Michigan, changing Earth magnetism surfaced as a primary factor in the choice of specific areas for field research in Wisconsin. In fact, magnetism in general became a mainstay in ongoing field research from that point onward.

Finally, inspection of magnetic maps for the area led me to believe that the entire Northern Highland geographic area of Wisconsin was wide-open for positive sasquatch exploration. We have certainly not been disappointed, because with 1,530,647 acres to wander in, clearly, bipedal anomalies would find suitable, longterm homes within this area. This early evaluation has certainly proved valid.

I used the *Total Magnetic Intensity Map of Northern Wisconsin*[28] as a reference for a magnetic baseline in the area; this allowed me to choose a logical camp area for our beginning Wisconsin research. The magnetic baseline in the Northern Highland geographic area is akin to that experienced in Michigan. No throwing darts at a map for us! Furthermore, recall that the bedrock formation known as the Laurentian Shield spreads across the whole of northern Wisconsin. It turned out that the Laurentian Shield is a valuable asset for sasquatch research—as you will continue to learn.

We did not have time during our spring reconnaissance trip to find a suitable base camp location so, in early August 2011, I phoned a decades-long friend who lives in the Watersmeet, Mich., area and asked him if he could drive to a zone I selected from a topographic map for a look-see. He kindly said yes. The zone I asked him to check was a nearly two-hour drive from his home, yet he willingly made the trip and reported back to me. I'll name him "WF" for Watersmeet Friend.

Again, it's vital to state that no research project occurs in a vacuum, and without the encouragement and support from many wonderful folks, this story would not be told. WF had hiked into the wilderness of the western Upper Peninsula of Michigan with me on many occasions over the early years of research. I shall always appreciate his many contributions to our overall goals.

At any rate, WF checked out the area and discovered that the campground of interest was no longer available. After driving around for a time, he landed on a spot he thought might work for use as a base camp. Later in August, Maggie and I, along with Brushtracker and Dragonfly, drove north where WF met us and introduced us to the area. Following discussion and much driving around, we finally landed on a base camp area we found off an old logging road not too far from our original area of choice. Here, we set up camp.

Refined Data Gathering?

Another important research change occurred in 2011; I began logging further pertinent data. For instance, due to the turn in my research approach and future goals, I logged weather, wind, temperature, average humidity, barometric pressure, and dew point. I also logged solar activity data, including solar flares, solar wind, planetary K index values—known as the Kp index—and the interplanetary magnetic field (IMF) components. I record only the Btotal and Bz values of the IMF because these proved the most germane to my research. I'll offer a brief definition here for the space weather values I chose to measure and record.

The *K index*, known as *Kp*, is a measure used to denote the strength of a geomagnetic storm. The National Oceanic Atmospheric Association (NOAA) sponsors the online Space Weather Prediction Center (SWPC), which rates geomagnetic storms from Kp 1-9, with 1 being the least severe and 9 the most severe. NOAA also rates these storms as G0-G5, where *G* stands for *geomagnetic*. It uses Kp indices 1-9 in correlation with the *G* designations to name and publish the strength of a geomagnetic storm. For instance, Kp indices 1-4 qualify as a G0 storm, Kp 5 as a G1 storm, Kp 6 as a G2 storm, Kp 7 as a G3 storm, Kp 8 as a G4 storm, and Kp 9 as a G5 storm, which is the most severe.

I also kept track of the IMF because this field plays a major role in how solar winds interact with Earth's magnetosphere. The IMF presents as a vector quality with three axes; two of these are parallel to the ecliptic and one runs perpendicular. My primary interest in the IMF derives from the perpendicular value, or the so-called *Bz* unit of the IMF. The Bz unit of the IMF is created by waves and other disturbances within the solar wind. The SWPC defines the Bz unit in their online glossary as:

> A measure of the North/South orientation of the interplanetary magnetic field [IMF] measured perpendicular to the ecliptic plane. When Bz is southward, or antiparallel to the earth's magnetic field, geomagnetic disturbances become much more severe than when Bz is northward.[29]

Therefore, when the Bz is in the *southward* direction, it is much easier for solar winds to enter Earth's magnetosphere and wreak havoc. Thus, as we log data, we look at not only the *numeric* value of the Bz unit, which is measured in units named nano-teslas (nT), we also record its *perpendicular* direction, be it north or south. That information provides a sense for potential geomagnetic influence within Earth's magnetosphere, which is central to our premises of how geomagnetic interferences interact with Earth's changing magnetism, humans, and sasquatches.

To address the question of why we are interested in geomagnetic interference within Earth's magnetosphere, we look to the effects such geomagnetic interference has on all things electrical or magnetic—including humans and all living organisms. In short, there are distinct types of space weather, with each type affecting different technologies on Earth. For example, SWPC named one type of space weather as Radio Blackout Storms, where Solar Energetic Particles (SEPs) can cause electrical failure when they penetrate electronic devices and equipment; these are rated R1-R5. Another one is named Solar Radiation Storms. These storms are known as Solar Proton Events (SPEs) and, according to *SpaceWeatherLive*, they "can bridge the Sun-Earth distance in as little time as 30 minutes and last for multiple days."[30] These storms are rated S1-S5. Finally, Coronal Mass Ejections (CMEs), cause a third type of storm, Geomagnetic Storms, and are rated G0-G9. According to the SWPC, each type of storm adversely affects technology differently.[31]

As interesting as that is, technology isn't the only impact that space weather has on Earth; humans and all living organisms are also affected when any of these storms hammer at Earth's magnetosphere. In fact, many studies over the past 15 years show that space weather causes changes to human health.[32] For example, some of the risks associated with a geomagnetic storm are heart rate fluctuations, heart attacks, strokes, acute coronary syndrome, blood pressure increase, seizure, migraine risk, anxiety, stress, emotional instability, cognitive diminution, suicide risk, mental disorder flare-ups, and radiation risks for airline passengers at high latitudes.[33]

Other risks include digestive disorders, skin conditions, visual impairment and emotional instability.[34] This isn't all. When the geomagnetic score (Kp index) is at zero for 12 consecutive hours, humans experience health risks from cosmic rays. Cosmic ray risks include acute myocardial infarction, cerebral stroke, terminal arrhythmia, anxiety, stress, emotional instability, cognitive diminution, uptick in traffic accidents, and mental disorder flare-ups.[35] My goodness! Who knew that space weather stemming from our Sun could be so varied, scientifically confirmed, and capable of altering human health? Amazing … and vitally important to our longterm sasquatch research efforts!

Next, let me offer a simple definition for several terms used in this book: magnetism, Earth's magnetic field, geomagnetism, biomagnetism, and solar storms.

Magnetism: The molecular properties common to magnets; the agency producing magnetic phenomena; the science dealing with magnetic phenomena.[36]

Earth's magnetic field: Earth's magnetic field (also known as Earth's magnetosphere or geomagnetism) is the magnetic field extending from Earth's inner core to where it meets the solar wind, a stream of energetic particles emanating from the sun. Earth's magnetic field protects it from cosmic rays that would strip away the upper atmosphere, including the ozone layer that protects Earth from harmful ultraviolet radiation.[37]

Biomagnetism: The magnetic field created by a living organism, the effect of an external magnetic field on living organisms, and the scientific study of these phenomena.[38]

Solar storms: The sun emits huge bursts of energy in the form of solar flares, coronal mass ejections (CMEs), or filament bursts. Each type of burst sends streams of electromagnetic energy toward the earth at high speeds.[39] Coronal holes also emit solar wind particles toward Earth. For instance, according to the founder of the online Space Weather News and author of *Weatherman's Guide to the Sun*, Ben

Davidson, "… coronal hole solar wind streams can cause auroras and geomagnetic storms at Earth."[40]

Okay. With that brief introduction on the data we collected, it's time to learn how we collect it. I use weather devices to gather weather data and refer to the online resource Weather Underground[41] for reference on historical weather data. I use spaceweather.com[42] for the solar data tabulated in our solar data tables. From this, you can guess that I log data in tables, one labeled weather and the other solar data. This allows for rapid data recording in the field and simplifies later report writing. I have tons of tables filled with countless data from years of research. That data supports most, if not all, the claims made as I share this story of sasquatch and me.

August 2011: Amazing New Data to Ponder

August 20: Setup Camp

Brushtracker and I decided to place two trail cams. Camera one faced the campsite and was placed 63 feet behind the back of my trailer, 30 feet high in a pin cherry tree, with the angle set to encompass 152 yards south-southwest from the tree.

Camera two was attached to the top of the roof access ladder on the motor home 14 feet above ground; its angle encompassed the backside of the motor home and trailer, pointing toward the woods. From the backside of the trailer to an area of grass and weeds was 22 feet, with 41 feet of wild raspberry bushes flourishing between there to the woods; thus, a total of 63 feet could be seen by the second camera.

Camera two snapped photo 4.1. Photo 4.1 qualifies as a blob photo; that is, some see a sasquatch in it, while others do not. If a sasquatch is in the photo it cannot be easily seen or readily picked out; therefore, the photo is known as a blob photo.

I present this photo here primarily as an example. I do not, however, include many so-called blob photos in this book because they are not definitive and tend to cause more controversy than they are worth. So, give photo 4.1 a brief glance and let it go. Unless, of course, you are driven to find the silly ole' *Wally* of old! Then … have fun!

Photo 4.1: Trail cam number two's view into the woods

August 21: Activity Noted at Camp

Imagine our surprise when we awoke on the morning of the 21st to a plethora of bipedal anomaly signs around camp. I was stunned, to say the least; ecstatic as well. Around 7:30 a.m. that morning, I walked around camp. Much to my astonishment, I noted **six** independent areas of activity that had obviously occurred overnight.

I'll begin by describing each of the six areas of activity, including four possible bedding spots. I named the activity zones activity areas 1-6. Activity area 1 presented with many footprints throughout, complete with an unusual circle at one end. The footprints were not castable; the site measured 15 feet 8 inches by 5 feet 7 inches and was 25 feet from the northwest edge of my trailer. The back edge of the trailer faced the side where I believe cinnamon fern and tall leafy spurge grew in abundance. See photo 4.2.

Photo 4.2: Activity area 1 — 08.21.2011

The unusual circle at one end of activity area 1 had a diameter of 3 feet 5 inches. The circle was unusual due, not only to its circular shape, but to the way the grass within the circle appeared to spiral in both clockwise and counterclockwise directions. Very odd indeed! Photo 4.3 shows a closer view of the circle in activity area one.

Photo 4.3: Activity area 1 circle

Activity area 2 offered a clear, albeit uncastable, footprint. So-called step and stride lengths were not measurable; again, due to the closeness of each print one to another. The print measured heel-to-toe 23½ inches, the mid-foot measured 5¼ inches, and the heel was 6½ inches. The print in photo 4.4 appeared 15 feet 4 inches from the northwest corner of the backside of the trailer.

Photo 4.4: Foot print in activity area 2

Activity area 3 presented as a 33-inch diameter circle, with the grass matted down in a seeming spiral fashion. The southwest edge of the circle measured 3 feet 5 inches from the north side of the trailer. A typical value for Earth magnetism in the overall area of research we were in measures at 50,000-53,500 nT (nanoteslas—nT), or 500-535 mG (milligauss—mG). The center of the circle measured 22,500 nT (225 mG), while just outside the circle measured 52,0000 nT (520 MG). Photo 4.5 shows activity area 3.

Photo 4.5: Activity area 3

This incredible difference in magnetic measurements within three feet is shocking, to say the least, and mind-blowing otherwise. The question here becomes, "What would cause such a drastic change of measurement between the circle and just outside it?"

I am not able to proffer a reasonable answer to this question. In fact, I couldn't answer it then and I'm unable to answer it now. Even so, as this story unfolds, I will later put forth a theory that might address such a measurement discrepancy.

Activity area 4 presented as a possible bedding site. It measured 13 inches at the head end, 45-inches wide at both the shoulders and knees, 39 inches wide at the elbow positions, and the overall length of the same impression measured 48 inches long. This impression appeared 13 feet 4 inches north of the trailer.

Measurements just inside each of the possible body positions—head, shoulders, elbows, and knees—was 20,900 nT (290 mG), while just outside of the clear impression, it measured 49,800 nT (490 mG). Once again, these measurements are inexplicable. Photos 4.6 and 4.7 show the impression from slightly different angles, giving differing perspectives on the impression seen in activity area 4.

Photo 4.6: Activity area 4

Photo 4.7: Another perspective of activity

Activity area 5 presented as another possible bedding site and seems quite similar to the impression seen in activity area 4. This one measured 12½ inches at the head, 44 inches wide at the shoulders and knees, 38 inches at the elbows and was 47 inches long. This impression appeared 11 feet 9 inches from the north side of the trailer.

Each of the possible body positions—head, shoulders, elbows, and knees—measured 21,200 nT (212 mG), while just outside of the impression the measurement was 52,200 nT, (523 mG). Once again, these are seeming incomprehensible measurements. Photo 4.8 shows the impression of activity area 5.

Photo 4.8: Impression of activity area 5

Interestingly, activity area 6 presented as a triangle. The length of the long sides of the triangle measured 27 feet 11 inches and the base of the triangle measured 6 feet 3 inches. This activity area appeared 4 feet 11 inches from the southeast corner of the motor home. Many uncastable prints were seen in activity area 5. Although we were unable to measure step or stride because the prints were too close together—rather like a shuffling walk with periodic standing—several of the prints measured as 23½ inches long, mid-foot at 5¼ inches wide, and the heel at 6½ inches wide. Do these measurements sound familiar? They should, for we saw identical measurements with the prints in activity area 2. Hmm!

Photo 4.9: Triangle of area 6

Both the base of the triangle and its apex measured 22,300 nT (223 mG), the center of the triangle measured 36,800 nT (368 mG), and anywhere outside the triangle measured 51,800 nT (518 mG). Photo 4.9 shows the triangle of activity area 6.

Clearly, August 21, 2011, turned out to be rather exhilarating. We spent the day snapping photos and taking measurements. Evening found us a tad or so befuddled and somewhat exasperated due to the mysterious aspect of some of the measurements. We could have talked ourselves hoarse, and nearly did, while desperately trying to explain the unfathomable.

Remember, these uncommon anomalies appeared during out **first** night in camp on our **first** field research trip into the Northern Highland geographic area of northern Wisconsin! We began our day with the highly unusual and ended it jabbering about the seeming inexplicable.

Reflect for a moment, on an earlier discussion in Chapter 3 about anomalous experience and cognitive dissonance. Perhaps this day presented us with validation that attention should be given to the vagaries sometimes associated with anomalous experiences. In this instance, the anomalous experience meant quite bizarre measurements. The inexplicable measurements combined with a total lack of understanding as to why these anomalies appeared overnight our first night in camp during our first foray into a new location of research baffled us. Indeed!

As this story of sasquatch and me unfolds into additional unexpected and, at times, totally unbelievable territory, I will return to these concepts. Why? Because repeated anomalous experiences permeate our research and the idea of psychophysiological responses rises over and over. Coherence, as mentioned in Chapter 3, offered a way to resolve some of the internal quandaries that became prevalent over the next few years. And once again, years of research confirmed that changing Earth magnetism does indeed affect the human body, and I strongly believe the same holds true for sasquatches.

Along a similar vein, I previously mentioned that a measurable energy projects outward from the physical body at least 10 feet. When this energy meets changing Earth magnetics, something happens. That is, as Earth's magnetic energy changes, the changes affect the energy projecting outward from the physical body. I propose that this, then, contributes to an overall result when one has an anomalous experience at the same time.

I might suggest that it is similar to when two or more numbers are added; the result produces a different quantity from the separate digits. Something similar occurs to the physical body when it undergoes changing Earth magnetics intermixed with an anomalous experience. These changes interact to produce a result; sometimes the result is extreme discomfort; other times, the result is extreme joy. The act of coherence alleviates such extremes, bringing balance to the overall experience. At any rate, the mysteries continued. Even with the notable activity and possible bedding areas, neither of the trail cams caught a clear photo of a sasquatch. Dang it! I sifted through nearly 6,900 photos and found very little worth mentioning.

There were a few blob photos; however, I won't insult you or me by reproducing more of them here. There were also quite a few interesting light anomaly photos, some of which seemed to have potential for depicting something; however, they did not show exactly what that something was, so I'll not place them here either.

Finally, Brushtracker and I noticed that a large branch had been snapped from the pin cherry tree where we placed camera one and tossed about 20 or so feet off to the side. We interpreted that as the sasquatches not wanting the cameras up—so we removed them and tucked them safely into my trailer. Hmm!

August 22: Kayak the local river

This day, Brushtracker and I kayaked a local river looking for signs that suggested bipedal anomaly activity. We were not prepared for in-depth exploration; rather, we intended to note and record specific locations for later investigation. We identified several spots that used sticks as markers of a sort; each of these areas had footprints coming from and returning to the nearby forest after a visit to the river.

Photo 4.10 shows one such stick arrangement that we saw. We spotted this one as we kayaked into the edge of the river bank. Careful observation of the riparian flora seen in the background reveals a trail leading to and from the water. Photo 4.11 is a close-up of the stick arrangement seen in photo 4.10.

Photo 4.10: View of stick arrangement at the river bank

To clarify, it is not uncommon in the realm of sasquatch research to see unusual stick arrangements. Oftentimes, a stick arrangement appears intentional, almost as if sticks were intentionally placed as a message of sorts. As we move through the sequence of the next few photos, I hope to show you what is meant by an arranged stick anomaly and how it might be interpreted as a form of communication.

In this instance, I'm convinced the stick arrangement was indeed intentional and the layout of the arrangement served as a message from one bipedal anomaly (probably an adult) to another bipedal anomaly (likely a young one). Hopefully, you will get the idea within the next few pages.

Photo 4.11: View of stick arrangement at the river bank

Photo 4.12 shows a view approximately 15 yards up from the river bank. Note in the upper left of photo 4.11 the appearance of a break on the branch nearest the front. This does not look like a beaver gnawed stick; rather, it appears as a relatively clean break. This branch has the identical diameter as the two branches lying atop one another with the longer branch facing toward the river as seen in photo 4.9.

It seems reasonable to deduce that these two branches were initially part of the branch in the upper left of photo 4.12 prior to being broken and placed near the river bank. The branch in the upper right corner of photo 4.12 does not, unfortunately, show the twist effect. I lost that effect while snapping the photo. Even so, you may note the bark peeling at the far right of the upper right corner branch, which resulted when that branch was twisted off.

Photo 4.12: Branches by the river taken from here

Next, a careful look at the grasses in front of the branches in photo 4.12 reveals what appears as footprints moving into the area. As well, note how the grass is matted down near the end of each of the branches that lost their tips. Again, one branch has a clean break while the other (not shown) was twisted off.

Photos 4.13 and 4.14 show what might be bedding areas for one adult and one young bipedal anomaly respectively; these areas were discovered 25 feet north of the broken branches seen in photo 4.12.

Photo 4.13: Adult bedding 25 ft. North of broken branches

Photo 4.14: Youth bedding 25 ft. North of broken branches

Finally, photo 4.15 shows two upright sticks stuck in the ground; these are seen in the very forefront of the photo, with a narrow diameter stick seen on each side of the forefront of the photo. Note how between the sticks and upward from the sticks, you might perceive footprints. These two sticks and the footprints led from the possible bedding areas to the river bank where the broken sticks are stacked atop one another as seen in photo 4.11.

Photo 4.15: Two sticks in the ground in forefront of photo

Now, then, if I were to conclude with a guesstimate, I suggest that an adult bipedal anomaly slept in one of the bedding areas and a young bipedal anomaly in the other. I believe it is quite likely that the adult placed the sticks in the fashion seen in photos 4.10 through 4.15 to help guide the young one safely to the water and back. This seems to align with a form of unspoken communication. Again, merely a guesstimate based on the overall arrangement of sticks, footprints, probable bedding areas, and obvious movement at this location.

Finally, we stopped at another location approximately one-half mile further north along the river and saw quite a few footprints and a trail leading from the river bank into the forest. Brushtracker followed the trail while I snapped photos of the evidence around the landing area because I was not dressed for the wilderness, merely a quiet kayak trip along a local river. He did not detect signs suggesting we pursue the trail further, so we headed back to the boat launch. None of the photos taken at the latter location were notable enough to post here for viewing.

We didn't do much research on August 22-23 because Brushtracker and Dragonfly were on vacation and needed time to play. However, I had a rather invigorating experience on August 24.

August 24: Hello, bear!

This day was essentially another blow-off day. WF popped into camp and he and Maggie hit the road to engage in one of her favorite pastimes: proving why she is known as the queen of garage sales. I, on the other hand, spent the day wandering the forest around the campsite to continue familiarizing myself with this new locale. The day was pleasant, and my time was well spent.

Maggie and WF returned early evening; I had one more area I wanted to check out before dark, so the two of them sat outside of the trailer while I headed back into the forest. I wasn't going far, no more than a couple hundred yards or so from camp. I had earlier seen signs in that area and wanted to double-check on whether they were worth pursuing. I arrived at the spot, which was a thick grove of young aspen complemented with a variety of brushes and tall grasses. Recall we were camped off an old logging road; hence, it wasn't unusual to see a grove of young aspens popping up after the loggers left.

There was a charming sugar maple tree gallantly growing amidst the young aspens. I sat down and leaned my back against the maple. Surely a delight to experience the serenity of the forest, with birds twittering (not *tweeting*), chipmunks scampering about—presumably happily—and the many delightful odors of the forest wafting gently past my nose. After about 10 minutes of a peaceful sit-down, I noticed the birds ceased chattering and the chipmunks became quiet.

Hmm! Something was moving through the brush just outside the aspen grove, and it seemed headed directly toward me. At first, I thought it might be WF looking for me. The sound got closer and closer until … of an abrupt sudden a black bear came to a complete standstill, looked directly at me, took a couple of whiffs through rather large nostrils, made a sharp 180-degree turn, and preemptively skedaddled out of there! Holy wow!

It is totally amazing to me how the bulky body of the bear moved in such a rapid yet graceful way. I estimated the bear at an average weight; somewhere between 225-275 pounds. I should have been terrified! Instead, I had the thought, "What the hell? Do I smell **that** badly?" I guess one never really knows how one is going to react in a situation like that until you find yourself in the middle of it.

I did, however, begin feeling somewhat apprehensive following my first silly thought. I removed my tape measure and measured the distance between me and where the bear did its acrobatic 180-degree turn to scamper away from me. **It was a mere 18 feet from me.** Okay, now my heart began beating a bit faster. Time to head back to camp, using, of course, a path **vastly** different from the one the bear took for its hasty flight out of the area.

Interestingly, when I arrived back at camp, Maggie and WF told me about the bear they had seen cross the camp area. Yup. It was headed in the precise direction I had found to sit and relax. That's all. No one was injured. And stuff in the wilderness does happen. I just wish the dang bear had run away from me after *seeing* me, not after taking a couple of *whiffs* of me!

After discussing the incident, we chatted a bit more about signs I had seen throughout the day as I wandered the forest. They also showed me some of the silly items they'd bought at yard sales. Goodness! Well, they had a great time and Maggie had her garage sale fix, which is always a good thing.

After WF left, Maggie and I chose to go to town for dinner. Unfortunately, we received notice of a family emergency and had to break camp and leave for home early the next day, thereby cutting this trip short. We did, however, return to the same base camp area in September 2011 to continue the research. Diagram 4.1 shows the path of the bear as it crossed the camp area while Maggie and WF were sitting outside the trailer.

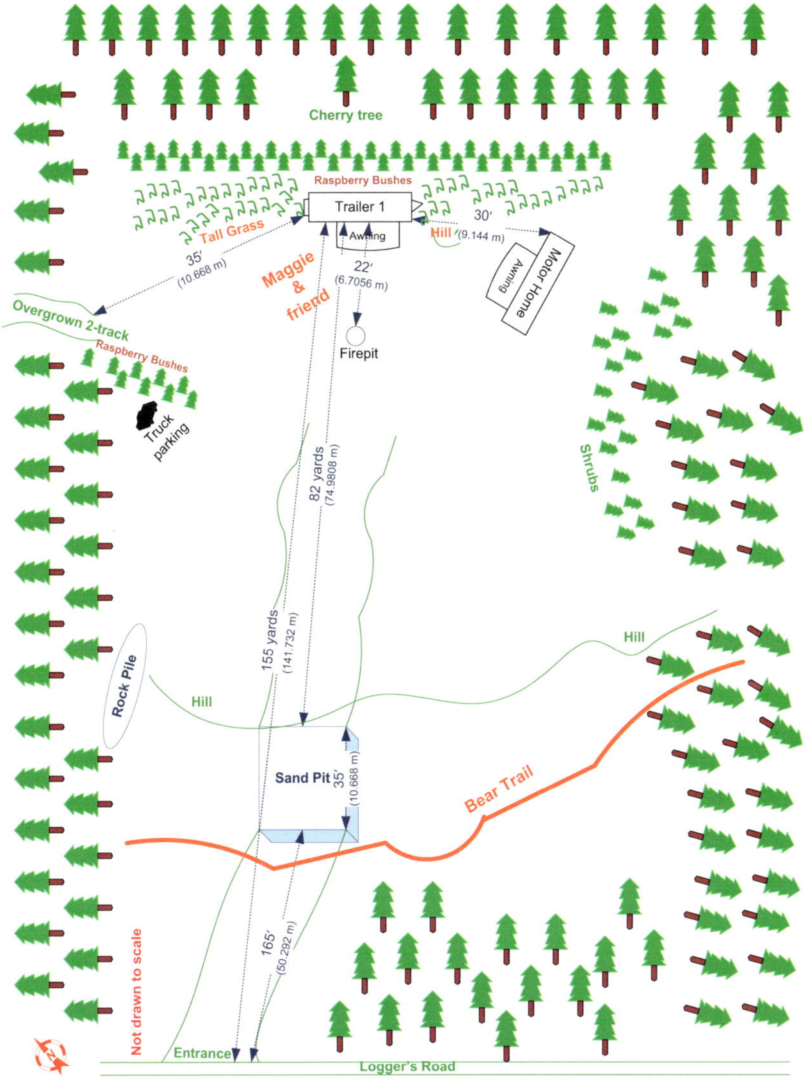

Diagram 4.1: August 2011 bear trail by camp

Chapter 4

[28] Karl, John H. "Total Magnetic Intensity Map of northern Wisconsin." Map. Madison: Wisconsin Geological and Natural History Survey, 1986.

[29] "Space Weather Glossary." Space Weather Prediction Center - NOAA. Accessed March 1, 2018. https://www.swpc.noaa.gov/content/space-weather-glossary#b.

[30] "What Is a Solar Radiation Storm?" SpaceWeatherLive.com. Accessed January 12, 2019. https://www.spaceweatherlive.com/en/help/what-is-a-solar-radiation-storm.

[31] "Space Weather Impacts." Space Weather Impacts. Accessed March 02, 2018. https://www.swpc.noaa.gov/impacts.

[32] Davidson, Ben. *Weatherman's Guide to the sun.* Place of Publication Not Identified: BookBaby, 2017, page 70

[33]Ibid.

[34] Ibid.

[35] Ibid.

[36] Magnetism. Dictionary.com. *Dictionary.com Unabridged.* Random House Inc. Accessed: March 7, 2018. http://www.dictionary.com/browse/magnetism.

[37] Dutoit, Yann Picand Dominique. "Earth's Magnetic Field." Earth's Magnetic Field : Definition of Earth's Magnetic Field and Synonyms of Earth's Magnetic Field (English). Accessed March 07, 2018. http://dictionary.sensagent.com/Earth's%20magnetic%20field/en-en/.

[38] *The American Heritage® Medical Dictionary.* S.v. "biomagnetism." Accessed March 7 2018. https://medical-dictionary.thefreedictionary.com/biomagnetism.

[39] Staff. "What Is a Solar Storm?" Wonderopolis. Accessed March 22, 2018. https://wonderopolis.org/wonder/what-is-a-solar-storm.

[40] Davidson, Ben. *Weatherman's Guide to the sun.* Place of Publication Not Identified: BookBaby, 2017, page 20.

---

[41] **Note:** Weather data accessed from: https://www.wunderground.com.

[42] **Note:** Solar data accessed from: http://www.spaceweather.com/archive.php?view=1&day=05&month=08&year=2011.

Chapter 5

A River of Surprises — September 2011

"When my information changes, I alter my conclusions. What do you do, sir [madame]?"

— John Maynard Keynes

The September 2011 trip had a different flavor to it; that is, vastly more time was spent in the wilderness. This is due essentially to the way I began to alter my research patterns. In other words, I continued to recognize a serious need to spend more alone time in the wilderness.

The sasquatch research had become an incredibly important part of my life and, as I conducted it through my business, I chose to treat it more as a business venture than a fun vacation or superfluous excursion with friends that included only short, essentially non-productive visits into the forest for serious research. This decision effectively turned everything around for me, albeit not right away. Why? At that time, I simply did not know how to make this happen without causing hurt feelings with friends or a serious ongoing debate with Maggie regarding her concerns about safety issues. Hence, while my desire remained strong to research alone, I felt conflicted enough to not make a final decision at that time. Instead, I continued to struggle with the issue.

Changes did begin to show themselves in the research, though, as we moved into and through the three trips north the following year in 2012. Each trip added confirmation to how important it was for me to be alone or with one close research friend. Yes, hindsight suggests I should have made a firm decision and acted upon it much sooner than I did. Alas, although I knew what I needed to do, I was seriously reticent to create waves amongst friends or with Maggie. Thus, I delayed acting on this decision. Not one of my better decision-making moments.

Amazing experiences did occur, though, once I finally made the decision to research alone or only with a close friend. You will see this unfold as this story moves forward.

September 2011 — A New Era of Magnetic Anomalies

September 22: Camp Set Up

Research was not conducted this day. Rather, this was camp set up time, complete with discussions on what I wished to accomplish during the trip.

September 23: Hiking Around the Camp Area

I had not yet explored the section of wilderness southwest from camp in this new area, so that afternoon I hiked in that direction looking for signs of activity of bipedal anomalies while familiarizing myself with the new location. It was overcast, and the light rain from the morning had stopped. I was excited walking about alone, for I had no distractions from others and was able to pay full attention to my wanderings as well as ponder the reasons I continued this research.

I walked a long half-circle from camp, thinking at the time that this would grant me the best opportunity to discover signs, simultaneous with getting a sense of the flora and topography of the new location. I had been out about 1½ hours and was at the apex of a large half-circle hike when the sky opened and dumped a chilly, driving rain onto the forest. Yikes! Not what I wanted. I quickly pulled my rain gear out of my backpack and practically jumped into it.

Once geared up, I checked my waterproof topographic map, looking for the straightest line from my current location back to camp. Wouldn't you know it? The straightest line from me to camp included a swamp area I had unwittingly avoided while hiking my long half-circle. Hence, the quickest way out of the forest was to continue my half-circle approach until I was beyond the swamp. I'll say this for northern Wisconsin—when the downpours start, they do so with great gusto.

Finally, after a couple hours of a slow, drenching hike, I had walked around the swamp and headed directly toward camp. The hike was slow because, in a rain-slick forest, the potential for accidents escalates and one needs to be poignantly aware of each step. It is better to be slow and soaked than to be quick, soaked, *and* injured from a fall. Another half-hour found me at the door of the trailer. Maggie, of course, had been a tad concerned and was relieved to see me.

I had seen many footprints and basic stick signs while hiking southwest of camp. The direction of movement for each of the prints was north-northwest. The direction of the prints suggested they would have gone past or near the trailer. Interesting. The so-called stick signs I had seen each pointed in the same general direction. The rain served as a severe obstacle to snapping photos, so very few of the photos I did manage to take turned out well. Irrespective, none were worth posting here.

Once I had dried off and changed into dry clothing, I pored over my maps and decided that when my friend Julie arrived to join us on September 24, we would start by hiking north-northwest of camp. Her joining me for my ongoing research became significant quite quickly, as you will soon see. Since then, Julie has spent more time in the field with me than anyone. As such, she has been deeply involved in this unfolding research for more than seven years. This becomes obvious as you continue reading.

September 24: Scouting the area for sasquatch signs

Julie arrived mid-morning and WF showed up as well. The four of us hiked into the area northwest of camp. Much to our surprise, we found interesting signs. For instance, we saw many footprints, quite a few possible stick signs, and an area we decided must be a potential bedding ground.

As this was a reconnaissance hike, I did not carry measuring equipment; an error I did not repeat during future reconnaissance hikes. We also saw fresh bear sign. Yes, fresh bear sign. The tracks of this bear were a bit larger than the one I had seen in August. We proceeded cautiously after becoming aware of the bear signs.

We also followed sasquatch footprints, one of which appears in photo 5.1. I estimated the stride of this bipedal anomaly at a little over five feet. After trailing these footprints for a bit, we found ourselves in an area we considered a group bedding ground. Fascinating.

Photo 5.1: Footprint

Only two of eight possible bedding sites appear in photos 5.2 and 5.3. The rest were similar in nature to these, and I believe that two are adequate to provide you with an idea of the way they appeared.

Photo 5.2: Possible youth bed

Photo 5.3: Possible adult bed

After WF left, Maggie, Julie, and I hiked back out for another look-see. Once again, we continued to find ample bipedal anomaly signs, so we split up and each took a bit different path while staying close enough to call out to each other. Throughout it all, we walked in a general north-northwest direction. I was on the far left, Julie the far right, and Maggie in the middle. After a short while, Maggie called out to inform me that Julie had found something she wanted me to see.

Sure enough, Julie had discovered a mesmerizing anomaly! A young maple tree had recently been bent over into a high arch. Wow! The young maple tree had a mixture of live green, red, and gold leaves on it; hence, our assumption of a fresh break. This was an amazing find for several reasons. First, it was freshly arched with the tip pointing in the general direction of travel for most of the footprints we saw earlier. Next, an enormous amount of power was needed to arch a tree this size. Finally, we deduced that at least **four** sasquatches contributed to the bending of this tree. Positively remarkable!

Photo 5.4: Break in trunk of the young maple tree (Maggie)

Note the break in the young maple tree from photos 5.4 and 5.5; now imagine the immense force needed to cause this break—the trunk of this maple had an 11½ inch diameter! The base of the young arched maple trunk presented with at least four, and possibly five, separate sets of footprints; hence, we deduced several bipedal anomalies were needed to bend this young maple tree into an arch. No small feat, eh?

How, I wonder, could we **not** be in awe of the extreme strength needed to bend that maple tree into an arch? And while the tree ripped in the center of the trunk, the roots were still strongly embedded in the ground. As well, the trunk was essentially intact—aside from the break area, of course. Incredible!

Photo 5.5: Julie and me under arched maple

Another thought-provoking factor; the top of the arch pointed in a north-northwest direction; the same direction in which we earlier saw a possible community bedding area. Hmm! Was this arched young maple tree a possible pointer toward the group bedding area? This was not a *stick* sign; rather, an amazing *tree* sign! Goodness! Guess we'll never know for sure what it meant to the sasquatches who made use of the arched tree; however, we can surely guess, eh?

We wandered a bit longer after snapping photos of the young arched maple tree and continued to see bipedal anomaly signs. While hiking, I couldn't rid myself of the feeling that I had inadvertently stumbled into a location with an absolute *wealth* of sasquatch signs. Without going all goofy mystic on you, I question at times if something known as synchronicity played a role in the pathway that eventually led me to this specific area within the Chequamegon-Nicolet National Forest. One thing for sure; the idea that this specific locale was worth exploring surely proves out as we move deeper into this story of sasquatch and me.

September 25: Amazing Kayak Trip

This day, Julie and I kayaked the same local river Brushtracker and I had during the August 2011 trip, only we kayaked in the opposite direction. The boat landing for the local river we were on this trip was about three-quarters of a mile or so from our campsite, so we didn't have far to go to begin our river excursion.

After paddling our kayaks northward for about three-quarters of an hour, we found a probable stopping point. We landed and began scouting around. Much to our astonishment, we found an amazing area that presented as a possible community watering hole. It had a well-used trail leading from the thick forest to the river and back where a noticeable number of footprints were seen. Further scouting produced another conceivable group bedding spot. Oh, my!

This remarkable discovery fueled our hopes and aspirations. In fact, these affirmations of major sasquatch activity in this area deepened our resolve to press forward. We realized we were blessed with dynamic proof that sasquatches—for whatever reason—gathered in large groups at times. Yet another amazing insight into the wiles of sasquatches!

This location was beyond interesting. We counted **16** bipedal anomaly bedding spots. **16**! Unreal. Furthermore, the bedding ground was in a half-circle, with the river fronting one side and thick forest surrounding the rest. The area facing the forest had twigs, sticks, and small branches placed around it about 15-18 feet from the bedding ground itself. We interpreted this as an intentional warning system to arouse those sleeping if something should approach while they slept. We also found footprints suggesting that a *guard* may have stood outside the circle of twigs, sticks, and small branches. The guard's spot is shown in photo 5.6; it stood just beyond the two branches seen in photo 5.6. Maybe the sticks and twigs were placed in case the *guard* dozed? Hmm!

This community bedding area rocked us into a bit of a jabber-fest. I mean, seriously? Was this bedding site for real? And what were we to do now that we'd seen it? We had been incredibly careful in our examination of it so as not to disturb the site with our prints; however, I suspect our smell alone would be obvious to whatever bipedal anomalies returned, if, indeed, this was not a single-night bedding ground.

Photo 5.6: Spot where a probable guard was standing

Of course, we also considered the unanswered question of why up to 16 bipedal anomalies were together in such a fashion in the first place. For example, does such a large group suggest migration? After all, it was late September and maybe they were moving to a completely new location for the upcoming winter. I guess they might also have been simply moving to a different locale within the region; again, due to the upcoming winter. Or maybe they gathered merely to enjoy a fall community dance by the river? This day was Sunday and maybe the fall dance had occurred on the Saturday evening prior? Hmm! Sorry we missed it … I think.

We didn't have answers to these questions then and we certainly don't have answers now. So much conjecture—so little actual knowledge. Nevertheless, using the zoom function on my camera, I snapped several photos of the bedding areas closest to me. One of these bedding areas appears in photo 5.7. The 16 bedding sites suggest that the creatures lie face down with their knees bent to their chests and foreheads touching the ground while sleeping. More on this later.

Allow me to reiterate, there were **16** separate bedding spots just like the one pictured in photo 5.7. Even more startling, finding this bedding ground occurred during our *second* trip to the Northern Highland geographical area. Yes, we were stunned. Yes, I remain astounded at the time of this writing.

Photo 5.8: Possible bedding spot

The following photo shows Julie peeking through the thickness of the brush surrounding the bedding area. This photo should give you an idea of how thick the brush around this area was at the time we were there.

Photo 5.7: Julie peeking through the brush

I've no clear idea how something as physically large as a bipedal anomaly can move quietly and gracefully through such dense areas. They do, though, and, I might add, quite successfully. We, on the other hand, seem to resemble a couple of bulls turned loose in a giant china shop. Break ... smash ... crunch—and this is when we are being stealthy! Oh, well.

We returned to camp and babbled to Maggie for quite some time about our day. She listened to us, tossing out comments and questions from time-to-time, and then viewed our photos. In the end, she said she wanted to see the area herself. Okay.

We made plans to take her out there on the morrow. Julie and I had kayaks and we needed another water conveyance, so we phoned WF and asked him if he could bring his canoe so he and Maggie could travel the river to the bedding spot with Julie and me. He said yes, agreed on a time, tagged off the line, and off to bed we all went.

September 26: Maggie and WF see the riverside bedding grounds

WF drove to our camp with his two-person canoe strapped atop his truck. WF, Maggie, Julie, and I hit the local river to paddle to the possible group bedding ground Julie and I had discovered the day before. Even though the weather was overcast, with a light rain, the exceptional beauty of the forest and the vibrant riparian flora along the river reached out to embrace us as we paddled. Delightful.

Much to the delight of my companions, I took a mild bath in the cool river water while attempting to get into my kayak. Hmm! Maybe I'll smell more like a sasquatch? Gosh! I hope not! I'll never forget the odiferous encounter of 1999.

Once Maggie and WF had a thorough look at the area, we stepped back into our water conveyances and wandered the river for a while. From time to time we stopped to assess areas at the riverbank and saw many signs; however, as we were on our way back to the landing, we did not invest much time in further exploration.

Photo 5.9 shows Maggie canoeing on the river—or at the very least, holding her paddle with a single hand. Impressive. Darn good thing WF was controlling the canoe, eh?

Photo 5.9: Maggie canoeing to bedding ground? Hmm!

Light rain continued to fall as we headed back to the boat launch. We stopped long enough for me to snap quick photos of a couple of possible bedding areas. One of these appears next as photo 5.10.

After arriving back at camp, we discussed the day's events, and WF headed home. We had completed a valued and beneficial research day, so we ate dinner, called the day to an end, and went to bed, not knowing, of course, what was in store for us the next day.

Photo 5.10: Rain enhanced photo of a bedding spot

September 27: Surprising Sasquatch Stick Sign

We awoke, had coffee, a quiet morning talk about our plans for the day, and then decided to take a short walk just north of camp. We left camp not knowing that, in a few short minutes, our world would not be rocked; rather, inundated with a tsunami of highly unexpected events. Please stay with me, here, because the next part of this story is quite extraordinary.

Our campsite was about 140 yards south of a little used east-west forest road. We walked through the woods north of camp to the forest road and looked across the road toward the area we had entered on the 24th, the day Julie first arrived.

I noticed a strange looking *something* in the ditch on the other side of the road. We walked over to it and goodness have mercy! It was a stick sign set there intentionally, and it dang well pointed **directly** toward our camp. Say what? How could this be?

See photos 5.11 and 5.12 for a look-see at the campsite pointing stick sign. Think what you will; these stick signs appeared overnight!

Photo 5.11: Campsite pointing stick sigh

Photo 5.12: Tripod base of the campsite stick pointer

Note how the base was constructed as a tripod with the pointer stuck through the top; the pointer was a straight stick with a slight natural bend at the tip. The tip protruded out from the top of the tripod base and pointed *directly* toward our campsite—I mean **directly**. One might think this was enough excitement for one day; however, one would be wrong to have such a silly thought. Much more was forthcoming.

After finding the campsite stick pointer, Julie and I checked out a meadow two-tenths of a mile from camp. This was a lovely, grassy meadow, albeit prone to flooding during the spring rainy season. See photo 5.13. This meadow gains *high* significance during a 2012 research trip, as you will see in Chapter 7. Until that amazing story unfolds, simply be aware that this meadow is not only lovely, it adds an important experience to this story.

Photo 5.13: West end of a meadow 0.2 miles from camp

Photo 5.13 shows the west end of the meadow. Julie and I walked the entire meadow looking for footprints. We saw quite a few and followed them into the forest at the east end of the meadow, and then walked north. We walked several hundred yards north of the meadow before we turned slightly northeast while passing through an enchanting and slightly open area blooming with copious moss and lichen, all sprinkled with a bit of Laurentian bulblet fern.

A couple hundred yards east, we discovered a very old, overgrown logging path that had large stones lining each side. While walking slowly along the path, we felt a sensation of being watched. I suspect that many reading this material are quite familiar with this commonly known sensation. After all, within the sasquatch phenomenon arena this feeling is considered an oft felt impression while plodding through a forest searching for sasquatch signs.

We found many footprints throughout the area and decided to take a few measurements. A specific stone stood out as somewhat significant. The magnetic measurements at this stone were reminiscent of the measurements taken at the campsite earlier this year during our August expedition, which merely added to the mystery of unusual magnetic measurements in this overall area. We noted the location on our map, knowing we would return when time allowed. That didn't happen, though, until the following year during one of our 2012 expeditions.

Having walked the entire stone-lined path, we found the eastern entrance to be interesting because of a small-diameter tree pulled into an arch directly over the path. Yup … another arched tree! Photo 5.14 shows Julie standing under the small-diameter arched tree. For future reference, we named this stone-lined pathway Dimension Lane (DL).

Photo 5.14: Julie & arched tree at entrance to DL

Longterm findings along Dimension Lane were extraordinary, as you will see once we skip into Chapter 12. For now, diagram 5.1 shows locations and measurements of six sets of stones lining each side of Dimension Lane. The stones, across from one another, are labeled V_{SSE} on the left side and V_{NNW} on the right side. SS stands for stone sets, SSE stands for south-southeast; NNW stands for north-northwest. The diagram shows six stone sets on each side.

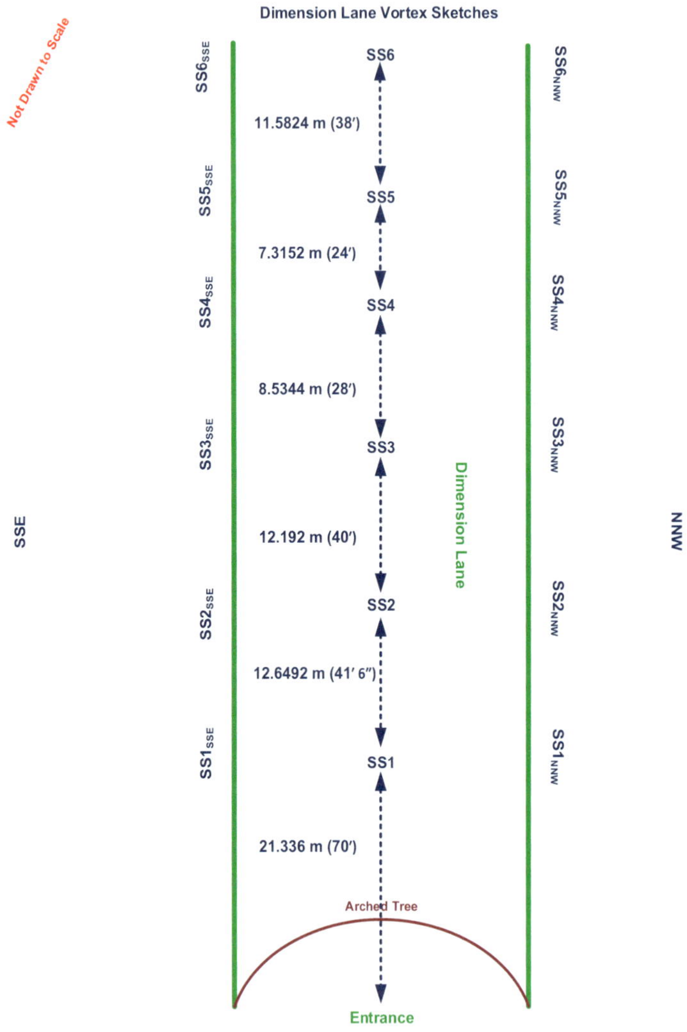

Diagram 5.1: Measurement between stone sets along DL

Our field time for this expedition was running out, so we recorded what we found, convinced we would return. The day was passing into early evening and we had quite a hike to get back to camp. As such, we left Dimension Lane—thoroughly happy with the discovery—and hiked out before darkness overtook us, chatting all the way about the unbelievable finds of the day.

September 28-29: Surprising Wrap-up of 2011

Julie left for home, so on September 28th, I took Maggie out for a look-see around Dimension Lane. She was as impressed as Julie and I had been. Because we had to break camp on the 29th, we wandered aimlessly for the rest of the day, driving around a bit to check out different areas. We saw beautiful country, and a plethora of bipedal anomaly signs everywhere we stopped. In the end, the width and breadth of the Chequamegon-Nicolet National Forest proved to be an area we'd spend the next few years exploring.

The many experiences that we had this trip left us feeling like tightly wound springs. Yet, there was one incident about to happen; one which drove me deeply, profoundly, and firmly into an unshakeable belief that sasquatches are real. On the evening of the 28th, we had an early dinner and then began preparing the trailer for the trip back home. We went to bed around 11:00 p.m. local time, which, in that area, is Central Standard Time. We chatted briefly, quietly about the potential that Dimension Lane offered for in-depth exploration, which wouldn't begin until 2012. Finally, we drifted off to sleep, hoping that dreams of continued success with our research would tweak our sleep with delight.

For some reason, I awoke suddenly, shaking off the sleepy confusion that often besets one waking abruptly in the middle of the night. I at once noticed an eerie dim light in the trailer, even though we do not make use of any sort of nightlights. The dim light allowed me to see the clock; it was precisely 3:00 a.m. on the 29th. The trailer window on the back side of the trailer is right next to my side of the bed and the blinds were fully open. I turned my head slightly … and was at once astonished into full wakefulness. Looking in the window was the face of a bipedal anomaly! It was bent over and looking sideways—directly into my astonished blue eyes! Was I in the proverbial Twilight Zone?

I was speechless, and momentarily motionless. Then, I slowly reached out my right hand to gently touch the window. The face withdrew a few inches, then returned. Its eyes were very large. They looked like deep, mahogany brown, and emanated a surprising intelligence. There was no hair on its face. The face was **huge**—larger than the window—and looked smooth like dark, tanned leather! It did not show its teeth; rather, merely watched me with what I perceived to be a questioning look.

The whites of the eyes were glowing, radiating a dim, whitish-yellow color. This then, must have been the origin of the dim light in the trailer. I reached over to wake Maggie; of course, by the time Maggie opened her eyes and woke up, the face was gone. Gone, I say! As if the whole affair had been part of a dream rather than real. Maggie listened to me, patted me on the arm, and exclaimed, "Wow!" She then went back to sleep. Damn! I know what I saw. It was real, it was huge, and its eyes showed an unexpected level of intelligence. An unforeseen treat and deeply touching. This experience left me KNOWING that sasquatches are real! Eventually, I fell back asleep.

Morning arrived, and while coffee was perking, I stepped outside for a look-see by the back window of the trailer. Naturally, my curiosity was heightened, if not exaggerated. Lo and behold, the hair on the back of my neck jumped to rigid attention from the goose bumps that assailed me. There were undeniable footprints just outside the window. It wasn't a dream! I had seen what I thought I had seen. Now my mind was all abuzz—and I hadn't had my first sip of caffeine. What now? How do I reconcile this experience with so-called reality?

Ah, hah! This then, is an excellent spot to elaborate a bit on cognitive dissonance and how it can affect our thought processes and actions. I'm almost sorry I brought the subject up because it is clearly a complex one. When confronted with seeming complex ideas or happenings, however, I often refer to them as simplistic complexity. This, because, although an idea or experience seems complex—and oftentimes is—I feel strongly that a simplistic approach can lessen the pain associated with wending one's way through whatever complexity a situation brings.

Recall that professionals suggest cognitive dissonance occurs when one is confronted with new information that contradicts … beliefs, ideals, and values.[43] Surely, seeing a bipedal anomaly's face in a trailer window two feet away from my face falls readily under this umbrella. As such, my experiences with so-called sasquatches persistently challenge my beliefs, ideals, and values. The question for me became "How in the world do I reconcile such challenges with my belief system?"

To start, I had earlier learned how to encourage a physical coherence between my physical brain and heart. A regular practice of coherence allows me to move into a state of clarity and calmness within a brief period. From a point of calmness, I'm then able to assess an anomalous experience with a greater sense of ease intermixed with clear thought. While this does not equate with not being surprised or shocked each time I see or interact with a bipedal anomaly, it certainly eases discomfort around the simmering sense of shock, or extreme heightened adrenalin rush, and leaves me able to interpret an experience with a greater sense of clarity.

In other words, I have an experience, get blown out of my mind, and then move into a state of coherence. This, in turn, lets me feel a sense of well-being psychologically, allowing me to acclimate to each experience with clarity of thought and emotional balance. From that state, I move into interpretation of an anomalous experience differently than I would if I had allowed cognitive dissonance to beset me with overwhelming doubt, disbelief, fear, or other negative emotions. There you have it. Simplistic complexity unraveled within a few intentional heartbeats!

Back to the proof. First, photo 5.15 shows the backside of the trailer with Maggie pointing at the window through which I saw the bipedal anomaly. Photo 5.16 shows footprints in a position where the bipedal anomaly was standing sideways to the trailer; that position allowed it to bend down and peek through the window.

The footprints were photographed near where Maggie is standing. Geez! A peeking bipedal anomaly! And that sucker's face was so large it didn't fit entirely in the window. Okay … what's next? Whatever it may be, we were definitely open to more surprises. I mean, like, who wouldn't be, eh?

Photo 5.15: Window through which a sasquatch peeked

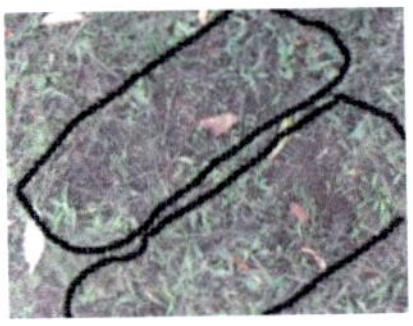

Photo 5.16: Footprints below the window

Well … is this a bunch of hooey or what! Not much to do about it now because today we drive back home. And believe it or not, this is just the beginning. Wait until we get into Chapters 12 and 13 where I relate details about the extended 2017 expedition, where I'm alone for nearly six weeks. At that point, the fire of disbelief really begins to crackle, spit and sizzle, and ultimately flames brightly into striking debate!

The window experience with a sasquatch face, then, is my **fourth** visual encounter with a sasquatch.

Chapter 5

[43] Wikipedia. "Cognitive Dissonance." *Wikipedia*, Wikimedia Foundation. Accessed February 22, 2018. en.wikipedia.org/wiki/Cognitive_dissonance.

Chapter 6

Reconnaissance in July — 2012

"Time spent in reconnaissance is seldom wasted."

—John Marsden

We did not take the trailer for our reconnaissance trip in July 2012; rather, we stayed in a cabin on a lake in a nearby village. From there, Julie and I drove 45-60 minutes one way to our intended research area. I mention this here because we were thrilled to learn that wild raspberries were bountiful and, we stopped each day on the way back to the lake cabin to pick a bunch. Adding fresh wild raspberries to vanilla ice cream for a nightly desert was exquisitely luscious. Yum! Now, though, on to the reconnaissance trip!

Hello have mercy! Our ongoing saga of sasquatch and me, magnetism, geomagnetism, and space weather continued when we stumbled onto yet another feature of magnetism during our July 2012 trip—that of crystal quartz. We found quartz in Dimension Lane (discovered in 2011) within many of the stones lining each side of the path.

Quartz is a crystallized form of silicon dioxide (SiO_2) and is used in many applications due to its silicon property. For example, quartz is used in watches, laser applications, and electronic devices of all kinds, including medical devices. Now then, we could go all crazy here and dive headlong into the scientific properties of quartz, such as examining diamagnetism, paramagnetism, ferromagnetism, piezoelectricity, induced magnetism, and so forth, and bore ourselves to death with ad nauseum technical detail. Or, we can treat the properties of quartz simply as a known electromagnetic quality in our research into bipedal anomalies. I prefer the latter for now. Later in the book, I will address some the above-mentioned properties of quartz. For now, though, let's just say *Yuk* to them.

First, all we need be aware of for now is that quartz embedded in stone creates an effect of some sort when it is excited by an external source, such as changing Earth magnetism, solar storms, or possibly an electrical storm when lightning and thunder combine to create a *vibration* within the quartz embedded in the stones along Dimension Lane. Moreover, nearby tapping or knocking sounds could also create a vibrational change of the electromagnetic qualities of quartz. More on this in Chapter 16. Potential magnetic effects induced by an external source can be measured with proper instrumentation.

When induced magnetism in quartz occurs, it produces the corresponding effect of something known as a resonant circuit, which includes a quality recognized as capacitance. Capacitance is an electrical phenomenon standing for the measure of ability for a given material to temporarily store energy in the form of an electrical charge. In effect, a "capacitor" can serve as a temporary battery or power source, producing the ability to momentarily power a device—or a resonant circuit.

This temporary charge in the form of electrical energy can be measured if the measurement occurs during the duration of the charge, or shortly thereafter while the provisional charge drains toward zero. In addition, crystals have an innate piezoelectric quality, which allows them to be used in frequency-determining circuits. One of the ways this can happen is in a so-called oscillatory circuit where oscillatory waves can be generated. I remind you here that Rollin McCraty stated that, "neural, chemical, electromagnetic, and oscillatory pressure wave patterns are among those used to encode and communicate biologically relevant information."[44] I will elaborate later on this idea. For now, we believe that the electromagnetic qualities of quartz found in the stones along both sides of Dimension Lane affected our magnetic measurements throughout 2012. There you are; simple, eh?

July 2012 Reconnaissance and Magnetic Anomalies

July 24-25: Dimension Lane & Triangle Village

Our visit to Dimension Lane on July 24 revealed an unusual find, which I'll get to momentarily. First, let me give you an overall impression of Dimension Lane. The total length of Dimension Lane, an overgrown incredibly old logging road, was measured at 566 feet 6 inches from the arched tree entrance to the end of the final stone set that offered unusual magnetic measurements. It's important to note, however, that the stone sets continued along the old, overgrown logging road; none of those stone sets, though, presented with unusual magnetic measurements past the 566 feet 6 inches that we defined as the length of Dimension Lane.

Dimension Lane ran in a general north-northwest direction from the eastern arched tree entrance. Along each side were stone sets of varying sizes. Many of these stone sets appeared directly or nearly across from one another. It was obvious to us that they had been in place for a long while. Moreover, close examination of them suggested the possibility that someone or something had at one time placed them in their current positions. Finally, many of the stone sets on both sides of Dimension Lane formed different-sized triangles. Interesting, to say the least, that many of the stone sets appeared as triangles of one type or another.

For your information, the last four pages of Chapter 7 show detailed diagrams of Dimension Lane. The diagrams include positioning of all the stones/rocks within each stone set along both sides of Dimension Lane. The final diagram includes a notation dubbed "DLT1," which we discuss in detail shortly.

The entire wilderness area surrounding Dimension Lane is comprised of thousands of acres of wooded land, with a liberal sprinkling of widespread stones and rocks, some of which were huge and others not so much. In fact, some of the rocks were so large it would take a platoon or company of bipedal anomalies to move them if such was their desire.

Dimension Lane, however, had 16 stone sets on the SSE side and 15 on the NNW side of the overgrown logging road for a total of 31 stone sets. Out of the 31 stone sets, 15 had stones placed in triangle shapes. Although these numbers and, most likely, the triangles carry no clear or obvious significance, the way the stones appeared does bear a bit of weight with respect to specific intent in laying them out. This, of course, is a guess based on observation and measurement.

At the northern end of Dimension Lane and on the northwest side, we found an area in 2011 we dubbed Dimension Lane Triangle 1, or "DLT1" for short. DLT1 turned out to be an area of primary interest during our July 2012 trip because of the way this triangle differed from our 2011 trip. Before we get to those details, though, let's first get a sense for the layout of DLT1.

Diagram 6.1 depicts the layout of DLT1 as seen on the afternoon of July 24. The diagram shows precise measurements; photo 6.1 shows the actual area. Unusual changes occurred within this area, both during our July reconnaissance trip as well as during the upcoming August and October 2012 trips. These changes over time caused yet more wonder on our parts.

As the changes occurred, we eventually gathered a sample of the soil from DLT1 to have it tested. A generous-hearted brother-in-law sent the soil to the Michigan State University Soil and Plant Nutrient Laboratory for evaluation.

Photo 6.1: DLT1; black circle is our area of interest

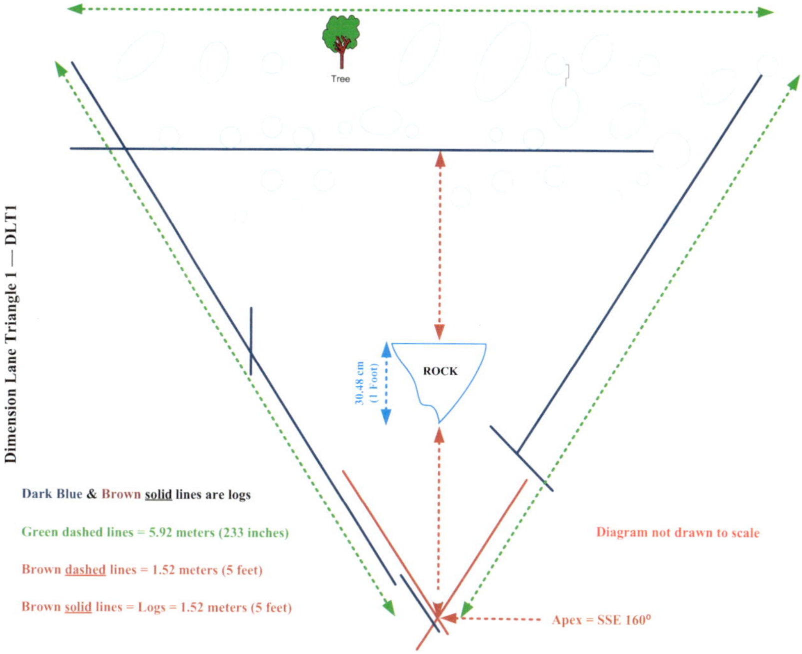

Diagram 6.1: Layout of DLT1 in Dimension Lane

Photo 6.2: Photo zoomed in for area of interest in DLT1

Photo 6.3: Apex of DLT1 as noted in diagram 6.1 as apex

Photo 6.3 shows a close-up of the apex. It seems obvious to me that the apex of DLT1 was designed and created with clear intent. Now let's zoom even closer into our area of interest in DLT1 to see what struck us as meaningful. Photo 6.4 depicts a zoom of the area seen in photo 6.2. The change here readily caught our attention because we detected scratches in the dirt. Photo 6.5 shows a close-up of *scratches* seen in the dirt.

Photo 6.4: A closer look shows scratch marks

Photo 6.5: Scratch marks; 13 inches high; 7 inches wide

All I want you to take away from the previous photos is that these scratch marks did **not** exist in August 2011. Furthermore, 13 inches high by 7 inches wide is a decent sized scratch mark! And this is only the beginning! Our August expedition is just around the corner, and "Katie, bar the door!" because you will see more major changes in this area from the way the photos display the scratches in DLT1 now.

After recording the above changes, Julie and I wandered back to the entrance of Dimension Lane and then continued south-southeast for about another 80-100 yards. Before long, we walked into an area that presented with four unique triangle-shaped sets of stones. We did a preliminary scout of the four triangle areas and, of course, decided to take measurements. The deciding factor for taking measurements was the bipedal anomaly footprints we found *within* each of the four triangles. Clearly, something was going on here; we just weren't sure what that was. We named this area Triangle Village, and the individual triangles as TVT1-4.

Photo 6.6: TVT3; Sasquatch footprints inside the triangle

Photo 6.7: TVT1; Sasquatch footprints inside the triangle

Photo 6.8: TVT2; Sasquatch footprints inside the triangle

Photo 6.9: TVT4 with arched tree; footprints under the arch

Photo 6.10: Crystal in rock

Photo 6.11: Closeup view of crystal in the rock

Crystals are present in the rocks found in Triangle Village as seen in photos 6.10 and 6.11. In fact, many of the stones and rocks in both Triangle Village, and Dimension Lane had crystals. Recall from an earlier mention in this chapter that we believe the crystals in these rocks played a role in the magnetic measurements we took throughout the area during our various expeditions to this zone.

As we continue with this story of sasquatch and me, I will have more to say about how we believe the qualities of quartz contributed to our resulting measurements. In the interim, simply recognize that I have thus far introduced several potential factors relating to magnetism and our search for bipedal anomalies.

From this chapter forward, I offer proposals that organize my beliefs about sasquatches as the research unfolds between 2012-2017. Here, I introduce the following two proposals:

Proposal 1:

> Changing Earth magnetism affects humans and all living organisms—including sasquatches.

Proposal 2:

> Solar and geomagnetic storms affect humans and all living organisms—including sasquatches.

I discuss much more on each of these thoughts as this saga continues. Our measurements at Triangle Village wrapped up our July 2012 reconnaissance trip. We concluded that we had a successful two-day reconnaissance excursion and were excited to continue our pursuit during the upcoming August trip.

Julie left, while Maggie and I stayed on for a couple of days to visit with friends before returning home. Lot's to discuss on the homeward bound drive!

Chapter 6

[44] McCraty, Rollin, et al. The Coherent Heart: Heart-Brain Interactions, Psychophysiological Coherence, and the Emergence of System-Wide Order. Institute of HeartMath, 2006, page 5.

Chapter 7

The Magic of August — 2012

"To succeed in life, you need three things: a wishbone, a backbone, and a funny bone."

— Reba McEntire

The August 2012 trip produced untold stimulating magnetic measurements. Our focus for this expedition was Dimension Lane, a thought-provoking area Julie and I had discovered in July 2011. This trip included Maggie, Brushtracker, Dragonfly, and me. Although I had not yet been able to quell my internal battle with myself over researching alone, I was moving closer to a clear decision on this and knew full well that the aftereffects from it would not be pleasant.

August 17: Set-Up Camp

We set-up camp, drove to Dimension Lane, and walked the length of it. Excitement reigned. We examined DLT1, taking a few measurements on it as well as the stone sets along Dimension Lane; these measurements served as baseline data for the trip.

Following our early examination of Dimension Lane, we agreed to return the following day, at which point we took *myriad* measurements.

These measurements greatly assisted my ability to assess where my research was and where I would take it in the future.

August 18: A Tedious Collection of Measurements

This day we focused on taking and recording detailed measurements throughout Dimension Lane. Brush-tacker and Dragonfly spent literally hours and hours helping us gather data on both sides of the full 566 foot 6 inch length of Dimension Lane. We are forever grateful for their dedication, patience, and support with this monotonous—yet vital—task.

Gathering quantifiable physical data and magnetic measurements in the middle of a wilderness is not necessarily the most entertaining way to spend a gorgeous August day in the Northern Highland geographical area of northern Wisconsin, especially if the time you take to do it cuts into vacation time. I was there to research; Brushtracker and Dragonfly were generously donating vacation time to aid our research. Maggie also wanted to play; as mentioned earlier, these considerations continued to weigh on me.

Nothing of note, however, occurs in strict isolation; rather, helping hands *always* play a role in a successful endeavor. Once again, without many helping hands along the way, this story would not be told. At any rate, while the data gathering was a rather repetitious task, the four of us managed to gather tons of data and have a bit of fun at the same time. In fact, I suspect we acted a bit like unbridled children at times; you know, giggling at the way a green frog jumped, or a butterfly that seemed attracted to Brushtracker! Oh, well.

We began our painstaking data gathering by measuring the distance from the entrance of Dimension Lane to a total of 31 stone sets—15 of which had triangle shapes. Again, the entire length of Dimension Lane for our August 2012 measurements was 566 feet 6 inches. One stone set was different enough to pay it special attention. I named the largest stone in that set Pyramid Rock. Pyramid Rock appeared as the 13th stone set on the south-southeast side and was 360 feet north-northwest from the arched tree entrance. This rock has an interesting shape as well as distinct colorations on its face, mostly from obvious weathering over the years. It is 5 feet 4 inches in height.

Photo 7.1 shows the front side of Pyramid Rock; this photo was taken the evening of August 17 during our early evening visit to Dimension Lane after setting up camp. Photos 7.2 and 7.3 show different angles of Pyramid Rock; one from the back, and one from the side. Stone Set 13 (SS13) had a total of six rocks, two large, three medium-sized, and one very large.

Pyramid Rock also proved to be the most magnetically active during our time studying Dimension Lane. I say this for two reasons. First, bipedal anomalies frequented this stone set more often than any of the other stone sets during our August 2012 excursion We found fresh footprints within this stone set *each time* we visited Dimension Lane. Second, this stone set *repeatedly* measured with the *lowest* magnetic values, which is highly significant. You will soon see what I mean by the latter statement.

Photo 7.1: Front of Pyramid Rock

Photo 7.3: Sideview of Pyramid Rock

Photo 7.2: Back of Pyramid Rock

To understand why the former statement is important, let me take a step backward and revisit an earlier research premise decreeing that changing Earth magnetism affects humans. As a reminder, typical values of Earth's magnetism in both the western Upper Peninsula of Michigan and in the Northern Highland geographic area fall within a range from about 49,500 to 52,500 nT (495-525 mG).

My earlier heart-head research proved that, as Earth's magnetism lowered below 40,000 nT (400 mG), automatic changes within the human body were recorded. For instance, when Earth's magnetism value dropped to 39,000 nT, blood pressure values began to *automatically* move toward normal regardless of whether they were low or high prior to this drop in Earth's magnetic energy. Furthermore, blood pressure value changes occurred with no external input other than subjects undergoing changing Earth magnetism. Heart rate values also began moving toward normal ranges when Earth's magnetic values were 39,000 nanoteslas and below.

Even more staggering, when Earth's magnetic values measured 28,000 nT and below, *all* checked bodily parameters moved *instantly* toward normal values, including heart rate variability (HRV). Please know this proves a profound and confounding anomaly: Earth's changing magnetism does indeed affect human physiological parameters regardless of location or baseline values! I address HRV more thoroughly in Chapter 15 because it relates to intentional coherence.

In the interim, while the above realization alone is enough to cause one to engage in stimulating discussion, I wish to acquaint you with the idea of coherence here. Much of my earlier research included a technique that the HeartMath Institute named coherence. Physical coherence, as measured by HeartMath in many clinical studies *stabilized emotions*, brought *mental clarity* to individuals and groups, and created an *overall sense of well-being.*

Stated simply, our earlier research about changing Earth magnetism repeatedly confirmed HeartMath's premises about coherence. In addition, our research showed that when subjects were connected to the HeartMath Freeze-Frame[45] unit and underwent changing Earth magnetism dips at, or somewhat below, 28,000 nT, they moved *at once* into a state of heart-brain coherence as displayed by the Freeze-Frame units. This occurred without engaging in intentional so-called balancing techniques—including intentional coherence—and a corresponding heart-head entrainment measurement was *immediate.* As an aside, I was able to connect four subjects at a time while conducting that research.

When changing Earth magnetism values reached around 25,000 nT, subjects also claimed to have a corresponding spiritual experience, where they felt a subjective and indescribable sense of unity with all around them. Frequently, the so-called spiritual experience included an inexplicable sense of joy, fullness, and completeness. I have years of data from earlier research expeditions supporting the previous statements. In addition, that research is reproducible today if undertaken using the same parameters adhered to at the time of the earlier research. To me, this stuff is fascinating as well as mind-boggling.

Now then, back to the seeming oddities of Dimension Lane. Many of the measurements made within Dimension Lane reached values as low as 35,000 nT (350 mG) and, at times, even a bit lower. Decades of reproducible clinical research allowed the HeartMath Institute to prove that magnetic energy *emanating from the physical heart* has been measured as extending outward at least 10 feet from the human body. Considering our earlier research, HeartMath's research, and data gathered at Dimension Lane, I arrived at the following proposal:

Proposal 3:

> Changing Earth magnetism has a noticeable effect upon the energy emanating outward from the physical heart—both for humans and sasquatches.

We spent an exhausting day cataloguing dozens of physical measurements along Dimension Lane and recording baseline magnetic measurements for the stone sets. These baseline measurements allowed us to watch for and record changes—if they occurred—throughout the rest of the trip.

In addition, when we first entered Dimension Lane this day, we noticed bipedal anomaly footprints at the arched tree entrance and they appeared quite fresh—almost as if a sasquatch was there either late during the night or early morning prior to our arrival. We measured the footprints as well as the *step*, and *stride* values. The footprints measured 21 inches long, $5\frac{1}{8}$ inches at midfoot, and $6\frac{1}{4}$ inches at the heel. As a point of interest, it is (at least for us) unusual to be able to obtain step and stride measurements; consequently, we were greatly pleased with this opportunity.

To define step and stride, I measure the **step** track from the toe of one step (the right foot) to the toe of the next step (left foot), as shown in diagram 7.1. The step track measured as 5 feet 4 inches. The **stride**, on the other hand, was measured from the toe of the right foot to the heel of the next right foot imprint. I measured the stride length as inches over eight clear strides (a total of 553 inches) to produce an *average stride length* of $69\frac{1}{8}$ inches, or about $5\frac{3}{4}$ feet. Diagram 7.1 shows the distinction between step and stride measurements. So you know, these are typical measurements when measuring step and stride values.

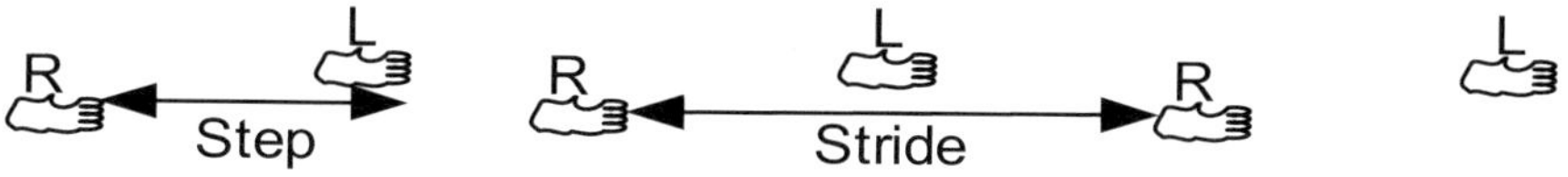

Diagram 7.1: Step and Stride measurements

The bipedal anomaly's footprints moved from north-northwest across the front of the arched tree entrance into SS1 on the south-southeast side. We have no idea how long it stayed within this stone set; only that it stood there for a time. From SS1, it moved south-southeast into the forest.

Diagram 7.2 shows the general path of the prints we measured. Again. It came from one side of Dimension Lane, walked across the old logging road, and then stepped into the area named stone set 1. Here it stopped for a bit, took another step, stopped again, and then wandered on into the forest.

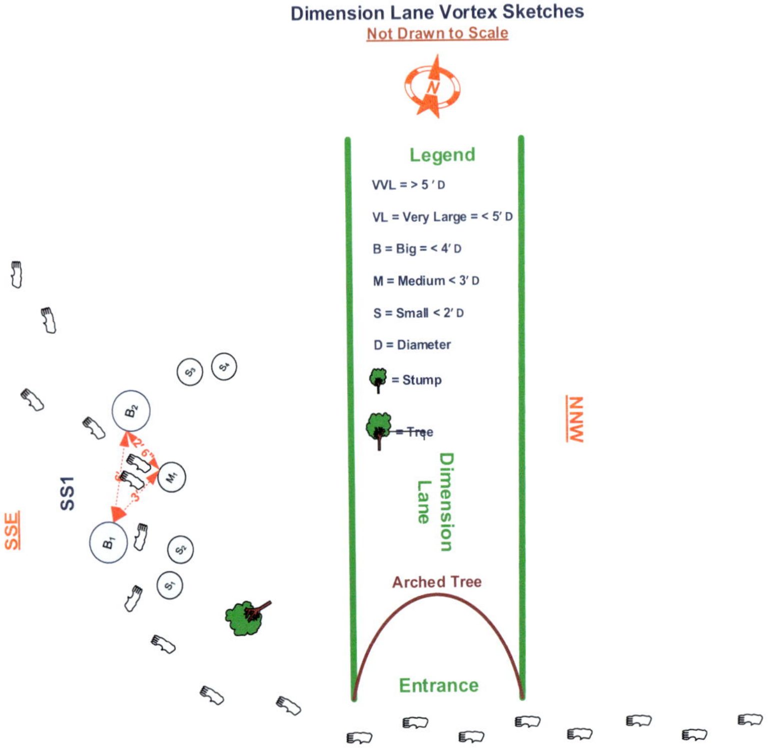

Diagram 7.2: Pathway of the prints seen on July 24, 2012

We did not follow the prints beyond about 50 yards after they left the general area of Dimension Lane. We were not interested in tracking the bipedal anomaly. Our interest was in measuring its step and stride data and pontificating on why it had stopped for a bit within the triangle of SS1 on the south-southeast side of Dimension Lane.

For purposes of later discussion, a magnetometer measurement of this triangular stone set, SS1, showed a value of 38,500 nT (385 mG). Its baseline value was taken the evening before and it read 44,960 nT (449.6 mG). This means that from the evening of August 17 to the morning of August 18 a drop of 6,460 nT occurred. Hmm!

Although not vastly overt, the drop is, nonetheless, meaningful. For example, three possible questions crop up here:

1. Did the magnetism lower *because* the bipedal anomaly stepped into the area?
2. Did the magnetism in SS1 lower, which then *enticed* the bipedal anomaly into it?
3. Or, perhaps, was there an immediate, mutual exchange of shared energy between the stone set and the sasquatch?

I do not have answers to these questions. I do, however, believe that once the bipedal anomaly stepped into this stone set, its magnetic energy interacted with the magnetic energy existing within SS1.

Regardless, we are sure of two things. First, the magnetic energy within this stone set was lower on the morning of August 18 then it was on the evening of August 17. Second, we know a bipedal anomaly stepped into this stone set and stood for a time before moving on. At this point, all else is speculation.

Photo 7.4 shows where the bipedal anomaly stood within the stone set. The sasquatch seems to have stopped once, moved a step, then stopped again. I deduce this based on the side-by-side prints appearing in *two* locations within SS1, as noted in photo 7.4.

Photo 7.4: Bipedal anomaly prints inside triangle

Because magnetometer values are discussed throughout the rest of this chapter, I include a table here depicting baseline magnetometer measurements taken on August 18 for all the stone sets along Dimension Lane. Stone sets are labelled SS1 through SS31 and noted as SSE for the south-southeast side and NNW for the north-northwest side.

Curiously, we noticed that each stone set with rocks placed in a triangle shape, regardless of which side of Dimension Lane they were on, measured lower than the stone sets without triangle formations. While I have thoughts about this measurement anomaly, this book is not the forum to present them because it would sidetrack us way too far and for far too long. Hence, I suggest you simply let your imagination take you where it will for now. I'm working on a website corresponding to this book, so maybe that will be the forum to address this specific measurement anomaly in-depth.

Table 7.1: Stone Sets Baseline Magnetometer Measurements

08.18.2012 Evening Measurements Used as Baseline Values		
Stone Set Number	**SSE Side**	**NNW Side**
SS1 (Triangle)	38,500 nT	
SS2	44970 nT	
SS3	43560 nT	
SS4 (Triangle)	38080 nT	
SS5 (Triangle)	38790 nT	
SS6 (Triangle)	39580 nT	
SS7	42090 nT	
SS8 (Triangle)	38260 nT	
SS9	50960 nT	
SS10	48730 nT	
SS11	44980 nT	

SS12	40870 nT	
SS13 (Pyramid Rock) (Triangle)	Bottom of Pyramid: 43670 nT Top of Pyramid: 40670 nT	
SS14 (Triangle)	37470 nT	
SS15	47060 nT	
SS16 (Triangle)	39190 nT	
SS17 (Triangle)		37000 nT
SS18 (Triangle)		36520 nT
SS19		49300 nT
SS20		49300 nT
SS21		48900 nT
SS22 (Triangle)		37900 nT
SS23 (Triangle)		394570 nT
SS24 (Triangle)		38600 nT
SS25		48956nT
SS26		52625nT
SS27		52115nT
SS28		50548nT
SS29		51228nT
SS30 (Triangle)		39651nT
SS31 (Triangle)		36798nT

Magnetic measurements were also taken at Dimension Lane Triangle 1 (DLT1) on the evening of August 17. These measurements served as baseline values for the rest of the August trip. Table 7.2 shows the baseline measurements recorded for DLT1.

Table 7.2: DLT 1 Baseline Magnetometer Measurements

August 17, 2012 Evening Baseline Measurements		
Location	**Inside**	**Outside**
DLT_1 Apex	54600 nT	54380 nT
DLT_1 NNW (Left)	53410 nT	55550 nT
DLT_1 SSE (Right)	52320 nT	54300 nT
DLT_1 North (Back)	56240 nT	55540 nT
DLT_1 Dig out	55930 nT	55930 nT

August 19: Where's the dirt?

This day, Brushtracker and I drove to Dimension Lane to take more measurements. Much to our wonder, the DLT1 triangle presented us with a startling change from the day before—dirt was missing from it! Say what?

Oh, my! First thing in the morning and beset by yet another mystery!

Definite activity had occurred overnight. For example, we discovered bipedal footprints in front of the apex of DLT1. Also, the sticks at the apex were in positions different from the previous day. Photo 7.5 shows the footprints and how the apex sticks were redistributed or otherwise disturbed. Compare photo 7.5 with photo 6.3 to see the difference between the two. Quite a change in appearance, eh?

We, of course, wondered if the prints we found here might be from the same bipedal anomaly whose prints we had measured the day before. We can't say for sure; however, these footprints measured the exact same as those from the previous day! That is, they measured 21 inches long, 5⅛ inches at midfoot, and 6¼ inches at the heel. Go figure. We believe it was the same bipedal anomaly; however, we cannot empirically prove it. Aw, heck! I'm going with the assumption that it was indeed the same one. So there!

Photo 7.5: Footprints in front of apex of DLT1 on 08.19.2012

Note in the middle of photo 7.6, and just inside the apex, that a slight hole appears in the dirt. This, then, is the area from which dirt was missing when Brushtracker and I arrived that morning.

Photo 7.6: Missing dirt area marked

Frankly, it appeared as if a goodly amount of dirt was missing. Close evaluation of the area around DLT1 showed no spots where the missing dirt had been piled. Fascinating! Researching bipedal anomalies surely presents one with curious and captivating phenomena, oftentimes when one least expects them.

Not only were there no nearby piles of dirt, Brushtracker and I walked a circular 100-yard perimeter looking for the missing dirt. Lo and behold, none was found! Hmm! This implied to me that the dirt was either eaten or transported to a location further than the 100-yard circular perimeter we had walked. Food for thought; which, of course, is not a new experience for sasquatch researchers. To the contrary, in fact. When researching sasquatches, the unusual seems to be far more prevalent than not, inciting a nearly constant activity of challenging conversations.

You may recall that the area from which dirt is missing was seen first in Chapter 6 in photos 6.1 through 6.3. You may further recollect a close-up of a *scratch mark* as seen in photos 6.4 and 6.5. Okay. Let's surmise for a bit. Julie and I saw and measured scratch marks in this area during our July 24-25, 2012 reconnaissance trip, albeit, no dirt was missing at that time.

This day, August 19 we discovered dirt missing from this area. Yesterday, Maggie, Brushtracker, Dragonfly, and I were in this area gathering myriad measurement data. We spent the entire day moving about, disrupting the natural setting, gabbing incessantly at times, tweaking each other's funny bones, and periodically annoying one another due to the tedium of our chosen task. In other words, we trampled all over the area encompassing Dimension Lane and DLT1, leaving enough human signs and smells to frustrate the most tolerant critter or bipedal anomaly. Clearly, we upset nature's natural state with our data gathering, surely leaving nasty lingering human scents from the day before.

Yet, lo and behold, the very next morning Brushtracker and I discovered at least one bipedal anomaly had visited DLT1 overnight. Apparently, human signs did not prevent a sasquatch from checking out the area overnight. I decided this was worthy of consideration, so slipped that piece of data into memory for later review.

Table 7.3 shows DLT1 magnetometer readings taken the morning of August 19. I shaded the table cells in table 7.3 where *drastic* changes in magnetometer values happened at the apex and within the digging area.

This data is really quite telling, and worth a second, third, or even a fourth look!

Table 7.3: Magnetic Readings Morning of August 19, 2012

August 19, 2012 Morning		
Location	**Inside**	**Outside**
DLT_1 Apex	**39600 nT**	51000 nT
DLT_1 NNW (Left)	50100 nT	52300 nT
DLT_1 SSE (Right)	49120 nT	54100 nT
DLT_1 North (Back)	49500 nT	52000 nT
DLT_1 Dig Area	**38250 nT**	**38250 nT**

A quick comparison between tables 7.2 and 7.3 show rather radical changes at the apex of DLT1 as well as within the dig area itself. In both instances, the values are significantly lower, suggesting a strong likelihood that whatever fussed with the apex and dig area did so during a possible changing Earth magnetism period. If so, I suggest whatever made those changes in the dirt *must* have been affected by changing Earth magnetism while moving the apex sticks around and while digging in the dirt—if indeed Earth's magnetism changed. That is, I have no way of knowing if Earth's magnetism changed in that spot or if, perhaps, a sasquatch somehow caused a magnetic change in the area. Hmm!

Note that the inside of the DLT1 apex dropped a bewildering 15,000 nT (150 mG). The DLT1 dig area (inside and outside) dropped a mystifying total of 17,680 nT (176.8 mG). I assure you, years of research data show drops of these magnitudes do **not** fall under the umbrella of typical changing Earth magnetism behavior. Rather, drops of these magnitudes exist **far outside** typical ranges. Bottom line here: This was yet another highly unusual occurrence of changing Earth magnetism involving a bipedal anomaly. I logged the information, and the question became, "What in the world do I do with it?" I continue seeking an answer to that question.

Next, from table 7.1, the baseline value of SS1 was 38,500 nT (385 mG). This means that between the evening of August 18 and this day, August 19, the magnetometer reading for SS1 *increased* by a staggering 13,000 nT (130 mG). The baseline value of SS13 measured 45,640 nT (456.4 mG), which means the magnetometer reading *decreased* considerably, dropping 8,440 nT (84.4 mG).

As a reminder, I can prove changing Earth magnetism affects physical body parameters of humans, and, once again, I propose the same is true for bipedal anomalies. Ergo, our goal at this point is to sort through magnetometer readings to figure out *when* changing Earth magnetism may have affected a bipedal anomaly. In short, the lower the value of a magnetometer reading, the higher the likelihood that a bipedal anomaly standing in that area at the time the magnetism changed to a lower value was affected by the changing Earth magnetism. Quite possibly, the converse may also have occurred. That is, a sasquatch may have caused the magnetism to change! No way right now to know for sure, eh?

To gain a clearer sense for the direction some of my thoughts are leaning, I later purport that a strong possibility exists for sasquatches having a skill-set allowing them to *intentionally* alter or otherwise affect changing Earth magnetism. Shut the front door! Have I gone completely batty? Or maybe I've slipped into a foggy realm of facile paranormal nonsense? Or ….

An answer of yes to either of the above questions could indeed hold a tiny grain of truth, although I would surely deny the latter part—albeit with a bit of a grin. I tender information later in the book lending support to the earlier (perhaps outrageous?) premise. My thoughts here, though, do depend on several assumptions. I name them below.

1. A bipedal anomaly's heart energy projects outward at least 10 feet, just as do humans.
2. A bipedal anomaly's anatomical structure includes a sympathetic and parasympathetic nervous system.
3. A bipedal anomaly can experience a form of coherence between the physical heart and brain, just as do humans.

4. Now for the 'kicker': It might indeed be possible for a bipedal anomaly to *direct* a state of physical coherence in a manner allowing it to interact with changing Earth magnetism in an *intentional, directed* way.

Returning momentarily to the changes detected at DLT1, the final pertinent piece of information for this day relates to it. Table 7.4 shows DLT1 dig depth changes. The depth of the dig at DLT1 measured 2½ inches on August 17; the depth measured a full 10 inches on August 23.

Each day the depth of the hole created by the missing dirt measured deeper, suggesting that *something* must have dug out more dirt each night. That same something also made the dirt disappear, either by hauling it away or eating it. I favor the latter suggestion, and as will be seen later, it turns out that I eventually produced evidence such dirt eating most likely occurred! Regardless, we were thrilled to be there while so many changes and variant sasquatch activities occurred.

Table 7.4: DLT1 Depth Changes During the August 2012 Trip

DLT1 — Depth Change	
Date	**Measurement**
08.17.2012	6.35 cm (2.5")
08.18.2012	~8. 9 cm (3.5")
08.19.2012	12.7 cm (5")
08.20.2012	15.24 cm (6")
08.21.2012	17.78 cm (7")
08.22.2012	21.59 cm (8.5")
08.23.2012	25.4 cm (10")

August 20: Intentional use of the Meadow seen in photo 5.15

In Chapter 5, I mentioned a meadow that was two-tenths of a mile from our campsite, and photo 5.15 showed the west side of that meadow. You may also recall that I said the meadow would play a significant role later. Well, now is later. On August 20, we returned to the meadow for a specific purpose: to place grape bunches in a predetermined location.

A pin cherry tree grew on the northeast end of that meadow. While most of the pin cherries were gone, many remained scattered about on lower branches. We found different sized bipedal anomaly footprints under the lower branches, suggesting perhaps, that a family of three bipedal anomalies frequented this hot pin cherry tree spot. I believe there were two adults and a young one. Having walked the meadow several times, we decided bipedal anomalies hung around the area on a regular basis. Hence, we chose a branch on the pin cherry tree from which to hang grape bunches.

We tied a total of 120 large red grapes in small bunches on a branch of the pin cherry tree (yes, we counted those suckers!). We wondered if a bipedal anomaly would stop by to check the pin cherry tree for cherries, see or smell the grapes, and then eat them. Our hope was that one or more bipedal anomalies would enjoy the grapes. I had read so-called *gifts* left for sasquatches are often not touched for at least three days. Because of this knowledge, we hung the grape bunches and left to engage in other activities.

The afternoon of August 20 found us in kayaks and a canoe paddling north on a local river. I wanted Brushtracker and Dragonfly to see the group bedding ground Julie and I had discovered during our September 2011 excursion (refer to Chapter 5, September 25). You might recall our level of excitement at finding the group bedding ground alongside the river; Maggie and I wanted to share the excitement we had felt then with Brushtracker and Dragonfly. This excursion, though, was not for research; rather, simply four folks sharing time together in the wilderness—again.

Darn it! Once we arrived at the location, it became instantaneously obvious that the group bedding ground was under water. Doggone it! Who gave the river permission to rise to this level? Well, we had a delightful paddling experience with friends, so the trip was certainly not a waste. We paddled in a dilly-dally manner on the way back to the boat landing, enjoying the riparian flora along the river as well as each other's company.

We stopped a few times at possible sites where it appeared bipedal anomalies had moved to the river for water. Nothing spectacular caught our attention, though, as we wandered slowly back to the boat landing. This activity concluded our outings for the day. We loaded the water conveyances into the truck and headed back to camp.

Later that evening, I heard the snuffling and low growling of a bear behind Maggie's and my trailer. It sounded as if it were about 25 or so yards beyond the edge of the forest. Maggie, Brushtracker, and Dragonfly were at our campfire while I had walked behind the trailer. I was sure that those three were not in danger as the bear seemed to be moving away from camp, so I quietly slipped back around to the front of the trailer and the campfire. That was the only notable experience we had that evening, although we believed we had heard a bipedal anomaly moving about in the woods much later in the evening. Anyway, after chatting for a spell about nothing and everything, we called it a night and retired to bed.

August 21: Back to Dimension Lane, DLT1

This day, Brushtracker and I once again started at Dimension Lane. We first checked DLT1 for changes. Holy scat! Where's the dang dirt? Yes, DLT1 once again had visitors overnight. And the dig area had altered a lot! You will note depth changes in photos 7.7 through 7.11.

This was an extraordinary phenomenon for us to observe as the depth of the dig altered substantively throughout the week. It was also my first experience with such an unusual happening. Clearly, this caught and kept my attention throughout the week. My goodness! Such an uncommon yet titillating experience; and certainly, one which tickled the thought processes and imagination all winter long.

Photo 7.7: More dirt missing from DLT1 … much more

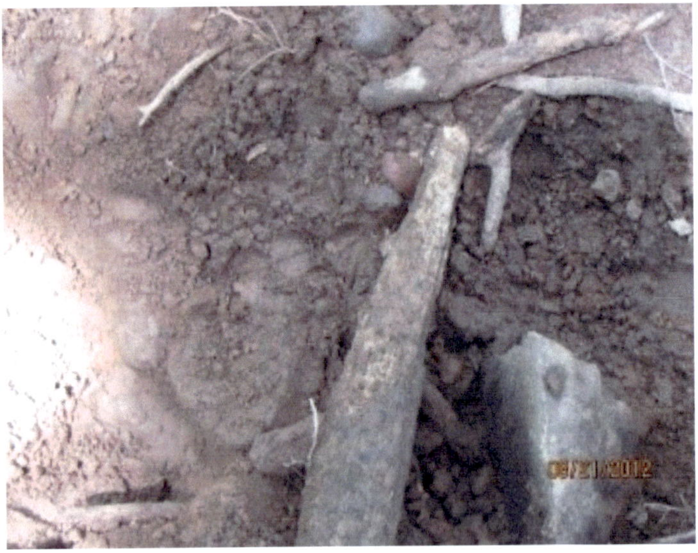

Photo 7.8: Circular impressions in bottom of dig area

Photo 7.9: Oh, my! Dig area expanded again

Photo 7.10: Side view; see rock leaning against the branch

Photo 7.11: Two prints nearly overlapping in front of DLT1

While snapping the above photos, Brushtracker and I heard a single bipedal anomaly engage in what folks in the sasquatch community call *wood knocking*. The sound came as a repetition of a specific pattern; that is, there were two short knocks, a brief pause, one knock, a brief pause, and two knocks. The wood knocking pattern repeated itself and lasted approximately 1½ minutes.

Then nothing; no movement sounds, no vocal sounds … nothing. We guessed the distance of the wood knocking sound at about 80 yards from us. Close enough to hear; too distant to see due to the denseness of the woods. We, did of course, eventually hike out to look for signs of the sound-maker. Goodness! Two grown adult men, whom we assume to be at least semi-intelligent, scamper off into the wilderness to find a wood knocking sound maker. Seems as if the older I get the more difficult it is to kick the little boy out of the grown man. Not sure I want to, though!

We did find footprints, and sure enough, the prints measured 21 inches long, 5⅛ inches at midfoot, and 6½ inches at the heel. Hmm! At this point, we automatically jumped to the conclusion that, although there may be more than one bipedal anomaly visiting Dimension Lane while we weren't there, at least one was dominant because we found prints with *identical* measurements in different locations on several consecutive days.

We confirmed our wood knocker was where we had thought. After looking around, though, we found no evidence of a stick or other object, such as two rocks, that might have been used to make the noise. Perhaps the bipedal anomaly carried off the sound maker ... or waved it like a magic wand and made it disappear? Whatever! Enough of Dimension Lane for this day.

That afternoon found Brushtracker, Dragonfly, Maggie, and I driving the area Julie and I had explored during July. I was interested in finding a different campsite.

I had good luck! I discovered a spot tucked further into the wilderness and with a lot less open space. My idea revolved around the thought that if bipedal anomalies wanted to visit us in camp, an enclosed forested spot would encourage them to come closer and perhaps, stay longer.

Maggie, Brushtracker, and Dragonfly were not impressed with the site. Even so, I made note of it on my maps because I was determined to use it as a future research basecamp. This site was about 20 miles from our original site, and I was thrilled to find it.

We checked the meadow once again on our way back to camp to see if the grapes had been eaten. Nope. Not yet. Little did I know at the time that once the grapes had been eaten, an incredibly amazing activity would follow. That excitement, however, appears later in this chapter.

August 22-23: Wander Time

August 22-23 did not entail research or data gathering; rather, these two days were spent wandering about in a lackadaisical manner as if we were tourists, not researchers. In the end, this August 2012 trip served as a striking catalyst for me to reexamine my strengthening need to not act as a vacationing tourist; rather, to engage in ongoing, serious sasquatch research. Hence, I needed to stop pandering around and deal with the issue of researching alone—or only with Julie—and address it in a candid manner. My internal conflictual struggle was wearing me out top to bottom and inside out, thus, impeding my research!

Again, I was not in the wilderness to play; rather, my goal was to engage in thoughtful, extensive sasquatch research. This festering issue had resurfaced, and rather than allow uncomfortable and building tensions to adversely affect a long-standing friendship, or to create marital conflicts, I made a final decision then and there that I would finally address the issue with Maggie, our children, and friends. Although I didn't know it at the time, this decision would ultimately reward me with astounding research experiences over the next few years, particularly in 2017. Of course, that same decision created ongoing tensions between me, family, and friends. Shoot!

Ah, yes. The so-called "butterfly effect" had finally taken wing! I had the thought and decided to confront the issue head-on. Even so, the effects of that decision would not come to complete fruition until 2017. Basically, that single decision on my part, though, set the stage for ultimate mind-boggling research results. I remain astounded to this day at the way future research activities played out following that single decision. Of course, I still needed to deal with expected repercussions. At least, though, I had made the decision to address the issue openly.

At any rate, Maggie and I continued to record changes to DLT1 throughout the week after Brushtracker and Dragonfly headed home. Daily visits produced further evidence something had dug dirt from DLT1, yet no dirt was found anywhere around the area. Just as Brushtracker and I had done, Maggie and I walked a 100-yard circle around the area and found nothing to suggest dirt had been moved to or piled in a location away from DLT1. I did, however, see another area similar to DLT1 that looked like dirt was removed from it. I did not, though, take time to monitor it like I had DLT1.

Imagine! The depth of the DLT1 dig measured 2½ inches on the first day we visited it. The last day's measurement was a gigantic 10-inch depth! That's a startling depth change from day one to day seven, a full 7½ inches. As noted earlier, I took a sample of the soil home for testing. A brother-in-law who is a wildlife biologist with the Michigan Department of Natural Resources (MI DNR) sent the soil sample to the Michigan State University Soil and Plant Nutrient Laboratory for me. I remain grateful to this day for his kind and generous aid. He is deeply appreciated!

The dirt was a reddish clay-type soil and test results showed potassium and magnesium as two of the ingredients. Let's take a moment now to briefly peek at potassium and determine why this alkali metal is significant to my sasquatch research. According to Wikipedia:

> Potassium is a chemical element with the symbol K (from Neo-Latin kalium) and atomic number 19…"[46]

In humans, the mineral potassium aids proper functioning of the heart, kidneys, and other organs. Once again, I assume sasquatches are carbon-based mammals; therefore, I suspect they require potassium for overall health and well-being, as do humans.

Further, Philip T. B. Starks, an associate professor of animal behavior at Tufts University in Medford, Mass., and Brittany Slabach, a graduate student, published an online article in *Scientific American* entitled, "Would You Like a Side of Dirt with That?" They wrote, "A common explanation for why animals and people eat dirt [geophagy] is that soil contains minerals, such as calcium, sodium and iron, which support energy production and other vital biological processes."[47] The missing dirt sample from DLT1 also contained magnesium and calcium. Apparently, sasquatches in this area were indeed engaged in a bit of geophagy. Yum?

Starks and Slabach also wrote:

> Today it is clear that geophagia is even more widespread in the animal kingdom than previously thought. Investigators have observed geophagia in more than 200 species of animals, including parrots, deer, elephants, bats, rabbits, baboons, gorillas and chimpanzees. Geophagia is also well documented in humans, with records dating to at least the time of Greek physician Hippocrates (460 B.C.).[48]

Considering that gorillas are included in the above quote about animal geophagia, it does not seem too big a stretch to me to suggest that sasquatches also likely engage in the practice in order to ingest supplemental nutrients they might not otherwise consume.

Another interesting tidbit from the Starks/Slabach study refers to the idea that geophagia might be used for *detoxification* purposes. For instance, they wrote:

> The idea that, in most cases, eating dirt is probably a way to get rid of toxins could explain why people and animals so often prefer claylike soils to other kinds of earth. Negatively charged clay molecules easily bind to positively charged toxins in the stomach and gut—preventing those toxins from entering the bloodstream by ferrying them through the intestines and out of the body in feces.[49]

A reference to dirt detoxification is also found in a 1986 article written by researcher Timothy Johns and published in the *Journal of Chemical Ecology*. Johns claims that dirt detoxification is an adaptive function of geophagy. He further suggests that:

> … geophagy [serves] as a solution to the impasse chemical deterrents pose … to chemical constraints on plant exploitation by non-fireusing hominids.[50]

In other words, some plants eaten by non-human hominids might contain poisons. Eating dirt—particularly clay—has been shown to detoxify chemicals from certain plants. Although this idea may not be terribly supportive of geophagy for the purpose of obtaining mineral supplements, internal chemical reactions do occur whether dirt is eaten for detoxification or as a supplement. Maybe a bit of both happens simultaneously for sasquatches? Conceivably!

Pearl Prihoda, a dedicated sasquatch researcher and author, addresses the importance of potassium as a supplement while positing her Photosynthetic Piezoelectric Induced Transparency (PPIT) theory. I address PPIT in Chapter 16. For now, Prihoda wrote, "potassium channels within our bodies are membrane associated proteins that function as electronic switches; when open, potassium usually flows out of the cell, altering the difference in electrical charge between its interior and exterior. Such ion movements enable nerve cells to send and process information."[51] My goodness! The importance of potassium to sasquatches can apparently not be understated. Too, I remind you that any external catalyst to any internal electrical quantity causes *magnetic changes*. The idea of magnetic changes continues to take on greater importance as we move deeper into this story of sasquatch and me.

Clearly, the dirt in DLT1 was missing, and we have convincing evidence for the presence of at least one bipedal anomaly in the area during the time the dirt was extracted. We also have a soil report showing that the dirt we sampled contained potassium (K-19), magnesium (Mg-12), and calcium (Ca-20). If we accept the above information about potassium—and include magnesium and calcium—it stands to reason that a sasquatch or sasquatches consumed the dirt. Quite likely for the potassium, magnesium, and calcium in it. Or, perhaps, due to the clay in the soil, a bit of detoxification might also have occurred. This conclusion might be akin to climbing out on a thin limb; however, I believe that while the limb might be swaying, it will certainly hold these thoughts, so I shall willingly climb and sway with them, at least until proven otherwise.

Now then, let's return to the grape bunches we hung on a branch of the pin cherry tree in the meadow two-tenths of a mile from our campsite. On August 22, the third day after hanging them, *the grapes were gone*. That is, all but 10 of the grapes were eaten; nine were left in one bunch and a single grape left in a second bunch. I've no clue why the remaining 10 grapes were not eaten with the rest. They looked fine to me. Photo 7.14 shows the ten remaining grapes along with the empty bunches. Photos 7.15 through 7.17 show footprints under the grape bunches.

As a point of interest, although you likely cannot see them in the photos, there are quite a few pin cherries left on the lower branches of the tree below the grapes. And, as a heads-up, the pin cherries still on the branches of the pin cherry tree become *highly significant* shortly. In fact, they are so significant that I suspect you will be shocked at the role they played in the end!

Photo 7.12: Grape bunches; 10 grapes left

Photo 7.13: One set of prints under the grapes

Photo 7.14: Another set of prints under grapes

Photo 7.15: A third set of prints under grapes

Maggie and I began breaking camp on August 23, preparing to leave for home the following morning. It was a quiet evening with just the two of us. We had a nice campfire and I began discussing with her my need to research alone or only with Julie. A vibrant conversation ensued!

We awoke early on August 24 and, while wrapping things up around camp, Maggie noticed something unusual and called me over to look at what she had found. As background information, I carry a large duffle bag filled with assorted sizes of wood I use to level the trailer, along with a few essential tools for quick access during the time we are camping. I had placed the large duffle bag just under the front corner of the trailer on the back (north) side facing the woods. I had occasion to access that bag several times throughout the week, then closed it and re-tucked it under the trailer.

Now for a strange yet startlingly cool story. Apparently, someone or something had seen me pull the bag out, unzip it, remove and replace something, rezip it, and then tuck it back under the corner of the trailer. Photo 7.18 shows the duffel bag pulled out from under the edge of the trailer.

Photo 7.16: Duffel bag under the edge of the trailer

Now let's walk our way through an extraordinary happening. Three days earlier, we had hung 120 grapes on a pin cherry tree branch in a meadow two-tenths of a mile from camp. We simply wondered if a bipedal anomaly might eat them. In the end, all the grapes but 10 were eaten. Of the 10 grapes left, there were nine on one bunch and a single grape on a second bunch.

The final night of our stay this trip, something appeared at camp with pits from pin cherries from the meadow. This day, the day we broke camp, Maggie found a goodly number of pin cherry pits in what I believe was a *fist indentation* at one corner of the duffel bag. Excuse me?

In short, I believe a bipedal anomaly left a *fist indent* in the edge of that bag. A total of 110 cherry pits were deposited in this fist indent. Photos of the indentation, pin cherry pits, and footprints were taken. Again, we had left 120 grapes in the meadow and 110 grapes were eaten; 110 pin cherry pits were left for us in a fist indent in the corner of the duffel bag! Seriously? Indeed!

Photo 7.19 shows pin cherry pits tucked into a corner of the duffel bag lying within an indentation. Following the photo, I offer my thoughts about this highly unusual occurrence.

Photo 7.17: Pin cherry pits in indentation of duffel bag

Please consider that for this to have occurred, something or someone had to engage in definite, well thought out steps of reason. For example, I suggest the following:

1. Over the course of a week, a bipedal anomaly or anomalies saw me accessing the duffle bag many times, thereby figuring out that the duffle bag was important to me.

2. The bipedal anomaly or anomalies either watched us hanging the grapes or smelled or sensed our presence on or around the grapes placed in the pin cherry tree. That is, a bipedal anomaly somehow recognized our scent and associated it with the humans camping two-tenths of a mile away.
3. One or more bipedal anomalies (I suspect three based on the prints) ate 110 grapes.
4. One or more bipedal anomalies ate 110 pin cherries, and *intentionally* saved the pits.
5. One or more bipedal anomalies carried 110 pin cherry pits two-tenths of a mile to our base campsite.
6. One or more bipedal anomalies had to find a place to leave the 110 pin cherry pits where we would find them.
7. The often-accessed duffle bag was selected to place the pits.
8. One of the bipedal anomalies had to—again, with specific intent—create a fist indentation in the end of the duffle bag.
9. Finally, the 110 pin cherry pits were then placed into the fist indentation for our later discovery.

Yikes! My goodness! Holy wow! This experience totally rocked Maggie and me to the core of our beings. Clearly, the above activities speak *strongly* of an ability by bipedal anomalies to engage in reasoning! At least, to a certain degree. Frankly, I get goose bumps just reliving that experience while writing it.

Goodness! Whatever species these bipedal anomalies turn out to be, those suckers can *think*—then act intelligently and with clear intent—upon their thoughts. Yes, I state that unequivocally. Furthermore, foolishly or not, I consider the pin cherry pits an amazing gift, a thank you, if you will, for us leaving the grapes. There are those within the sasquatch community who know what I mean; they've felt something similar. To those folks, I simply say, "How totally uplifting is this?"

Well, we finally got over our amazement, completed the breakdown of camp, and drove home. Maggie and I truly enjoy our driving times because of the opportunity for lengthy and lively communication. This trip home surely filled the cab of the truck with conversations concerning the many amazing events that unfolded throughout the previous week! In addition, I raised the issue of my internal struggle regarding my need to research alone. This conversation was not as pleasant as the earlier ones. Even so, I opened the door and the breezes of my need to research alone kept that door propped open.

Although it would be a bit longer before I took complete control of this need, when I finally did so, the results were phenomenal. After many amazing experiences during my nearly six-week stay in 2017, mostly alone, I no longer harbored regrets for eventually having made the decision to research alone.

Events of this week caused me to create another proposal. I base this proposal primarily on the actions associated with receiving 110 pin cherry pits from a sasquatch.

Proposal 4:

> The sasquatches with whom I interact show an ability to think and employ an early form of reasoning.

While I might be inclined to offer a specific type of reasoning, I shan't do so here. Gosh! The whole idea of reasoning—along with various types of reason—is enough to give a stout, healthy tree a serious root-ache! Imagine what it does to a human brain. Sure, and I'm glad I've learned how to engage in coherence exercises! Well, I'll list four types of reason and let you choose which one you think applies in this instance. Just do not ask me to clarify your thoughts for you because I—at times—struggle to do that for myself! Four types of reason are: (1) Deductive, (2) Inductive, (3) Abductive, and (4) Inferential. Have fun!

In closing Chapter 7, the following four pages of diagrams give you a comprehensive visual of the layout of the stone sets discovered within Dimension Lane. I feel it important to place these here so it's easier to follow this wild story. We aren't finished with Dimension Lane, as you will see when you dive into Chapter 8.

Diagrams of Dimension Lane and DLT1

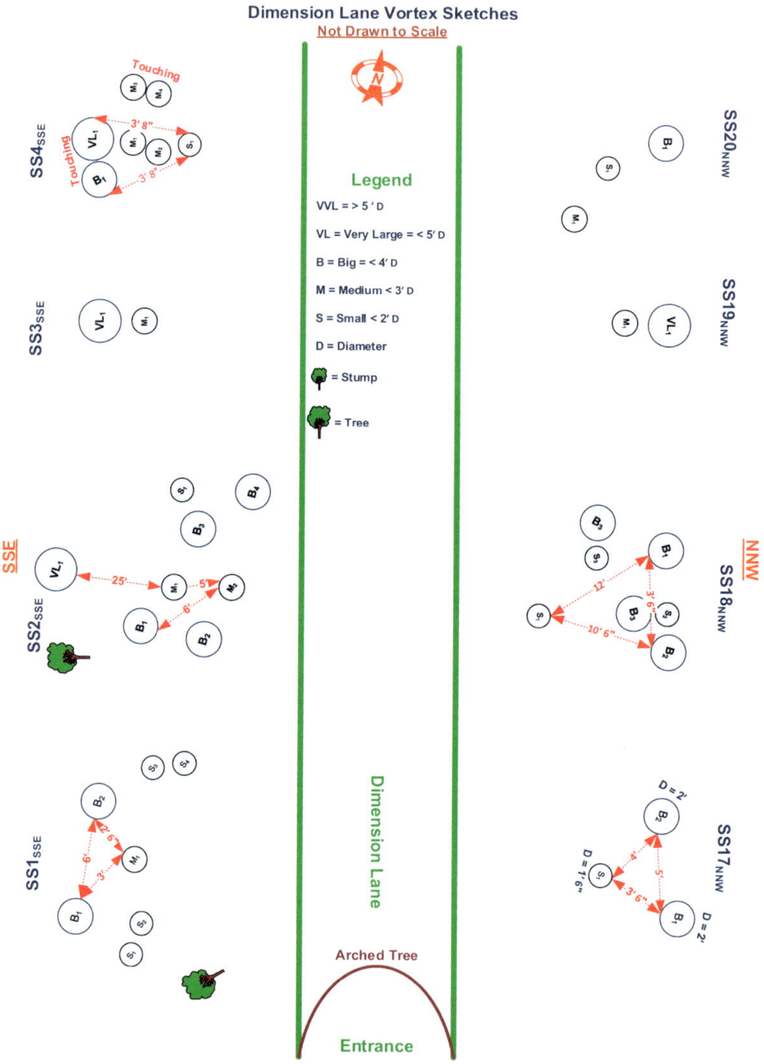

Diagram 7.3: Entrance to Dimension Lane

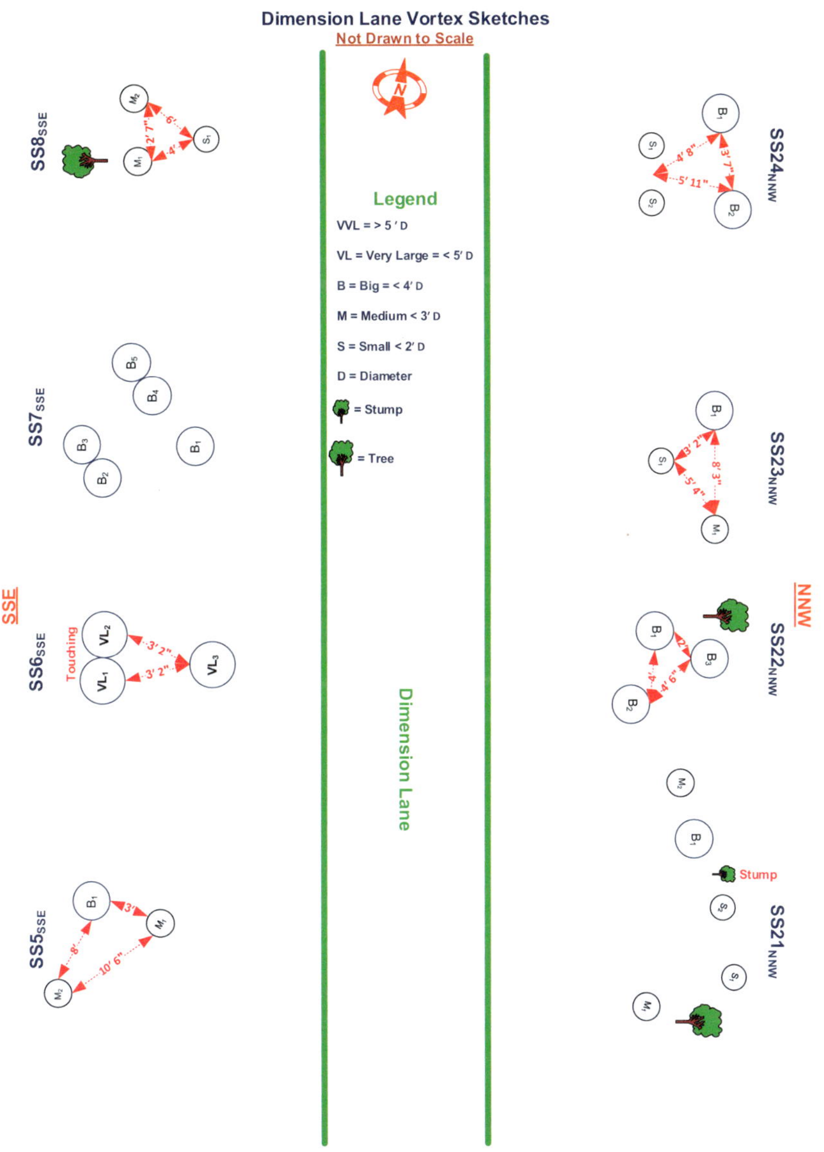

Diagram 7.4: Second leg of Dimension Lane

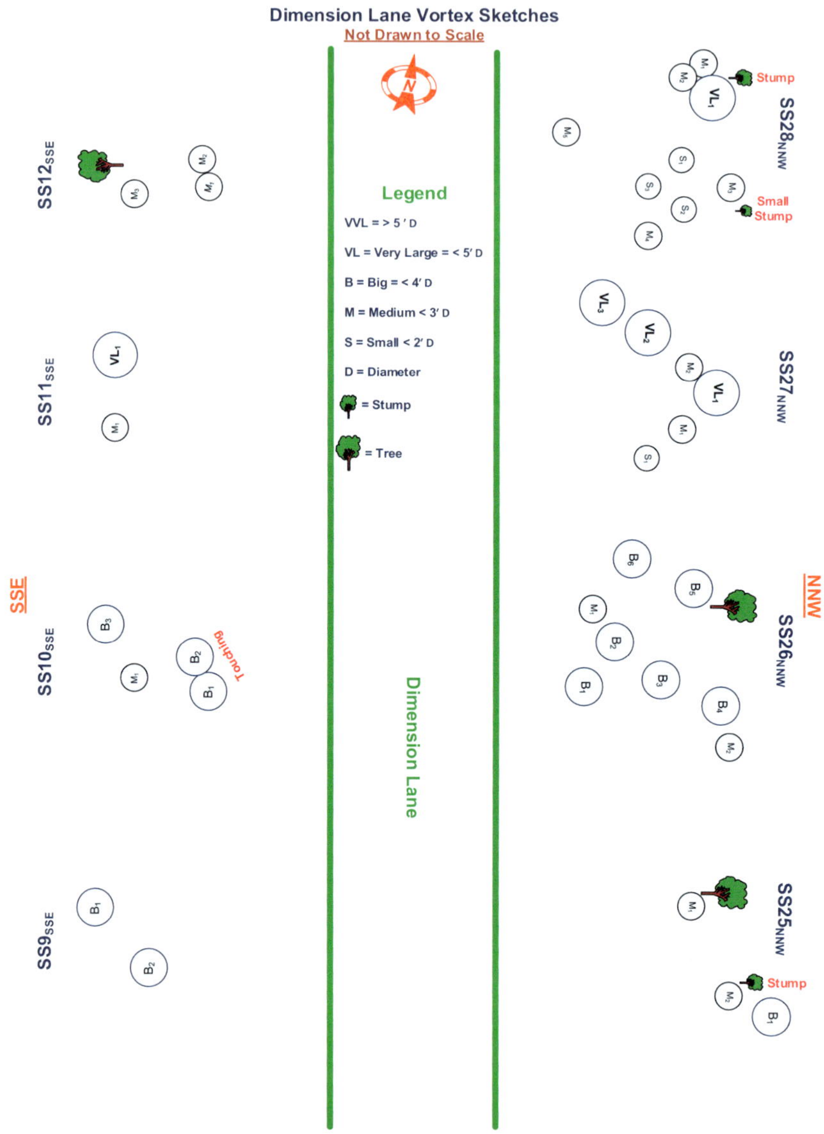

Diagram 7.5: Third leg of Dimension Lane

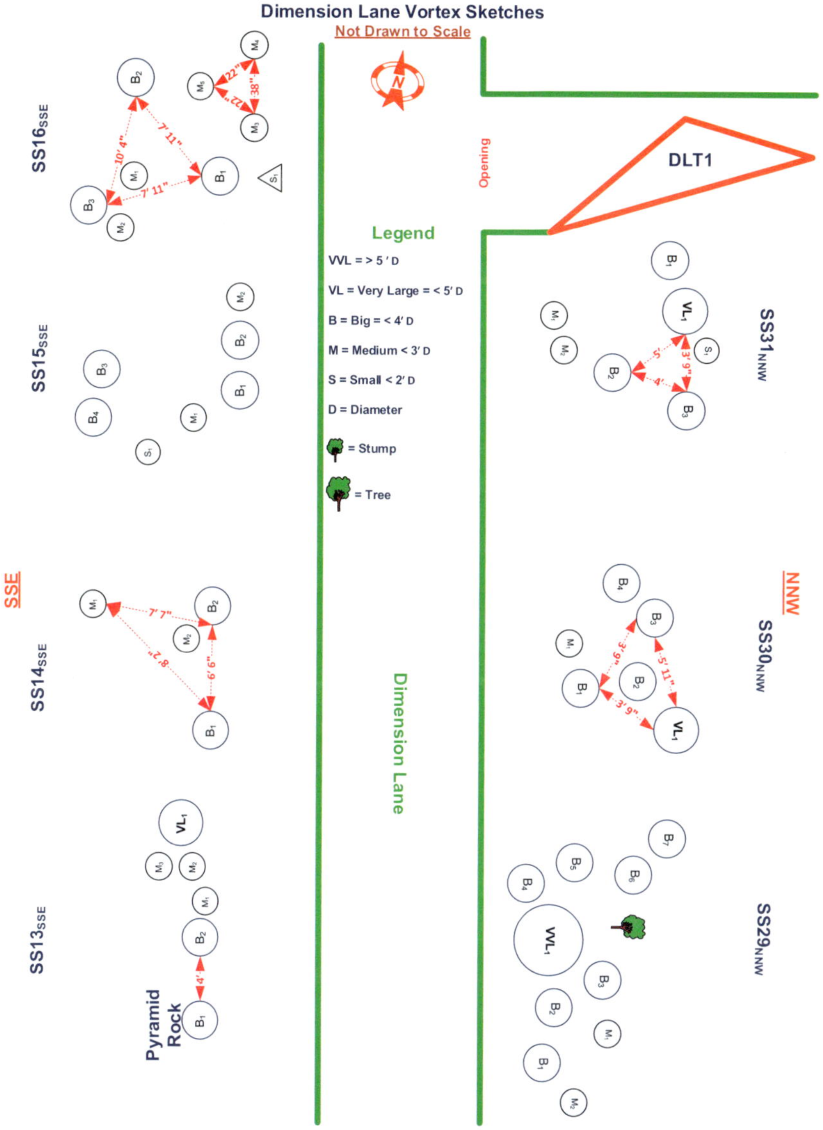

Diagram 7.6: Last leg of Dimension Lane

Chapter 7

[45] Freeze-Frame is a registered trademark of the Institute of HeartMath.

[46] Wikipedia contributors, "Potassium," *Wikipedia, The Free Encyclopedia.* accessed 11 November 2017. https://en.wikipedia.org/w/index.php?title=Potassium&oldid=809608643.

[47] Scientific American, "Would You Like a Side of Dirt with That?", Phillip T. B. Starks and Brittany L. Slabach, 01 June 2012. https://www.scientificamerican.com/article/would-you-like-side-dirt-eating-soil/. Accessed 11 November 2017.

[48] Ibid.

[49] Ibid.

[50] Johns, Timothy. "Detoxification Function of Geophagy and Domestication of the Potato." *Journal of Chemical Ecology*12, no. 3 (March 1986): 635-46. doi:10.1007/bf01012098. Accessed 11 November 2017.

[51] Prihoda, Pearl J. *Man-Otang: Bigfoot Myth or Reality*. Bellevue, WA: SAFEHouse Publishing, 2014, 67.

Chapter 8

The Wonder of October — 2012

"… the only way to do great work is to love what you do."

— Steve Jobs

October 2012 was the first truly fall research trip we took to the Northern Highland geographical area in Wisconsin. We were a bit or so taken aback to see the forest we had tramped through so many times over the years devoid of leaves. My goodness! Such a difference. It not only looked bare, it felt naked, open, and less obliging than when in full bloom.

I'm not sure why it felt this way; only that it did—at least, at first. Lesson? Do not let initial impressions drive your reality. Moving right along…

This trip found us staying with WF near Watersmeet, Mich., and driving daily (1½ hours each way) to the research location. Frankly, this was uncomfortable. First, I had become accustomed to interacting with bipedal anomalies when they visited our campsite. Second, Maggie enjoyed spending time with our friends, diminishing our research time. Once again, the issue for me was, "How important is this research to me, and what do I need to do to make sure the research comes first without hurting feelings or pissing off friends or family?" Clearly, I needed to address this issue as considerately as possible! So much frustration over feeling constantly conflicted, even though I knew what I needed to do. I simply didn't want to make waves. I did, however, get over that in time and began researching the way I needed to.

October 15: Afternoon Check of Dimension Lane

This day found Maggie, Julie, and I checking out the now renowned Dimension Lane for changes that occurred since our August 2012 trip. Upon arriving at the location, we walked the entire length looking for changes.

We did not see fresh prints or signs of other activity along the length of Dimension Lane. DLT1 did not disappoint us, though, for it showed a *major* change. See photos 8.1 and 8.2 for chronicled changes.

Photo 8.1: DLT1 dig area change from August 2012 trip

Photo 8.2: DLT1; Notice the leaves in DLT1

A hole had been dug into the center of the DLT1 dig area. Shining a flashlight into the hole showed nothing; simply dirt. No tracks of small animals; no acorns or other indications of the hole being used as storage for winter food. Nothing other than dirt. There were no fresh prints and, as you can see, leaves had blown into the area, suggesting nothing had been here for a while. It looked as if the attraction of DLT1 to sasquatches had waned; either that or the sasquatches decided to move elsewhere due to the upcoming winter.

A quick walk about the area confirmed our suspicions that Dimension Lane no longer served the purpose it had earlier. We did spot an interesting root-bed of a downed tree that either had gone unnoticed during earlier trips or had fallen since we were last here; my bet is on the latter because it is not unusual for forest trees to either uproot or break during high wind or electrical storms. This downed tree measured 198 feet north-northwest from DLT1. It looked as if a bipedal anomaly had used it as a bedding spot; see photos 8.3 and 8.4.

Photo 8.3: Root-bed of downed tree; possible bedding spot

Photo 8.3 shows the root-bed, while photo 8.4 shows a close-up of the interior. Note in photo 8.4 that an *impression* left on the ground looks a lot like that of a sleeping bipedal anomaly.

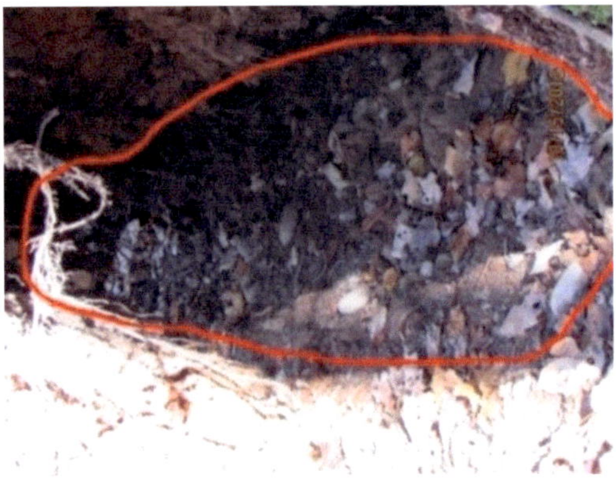

Photo 8.4: Possible bedding site under root-bed

Time to back-track a bit. I may have been remiss earlier in the book by not adequately describing the most common theory of how a sasquatch is thought to sleep. I'll rectify that now.

One of the well-known photos of a sleeping sasquatch originated with a Canadian, Adrian Erikson, and the then Erikson Project.[52] While much controversy surrounded his work, as it typically does with anyone engaged in sasquatch research, the project produced a photo entitled *Sleeping Almasty*.

The Erikson project was allegedly bought out circa 2012 by another controversial figure in the sasquatch arena, Melba Ketchum,[53] who is credited with having performed DNA sequencing on sasquatch hair, feces, etc. Nevertheless, the famous Erikson project photo has appeared many times on various online sites, resulting in it becoming well-known within the sasquatch community.[54] Although I do not reproduce that photo here, diagram 8.1 shows a rough sketch of a so-called sleeping sasquatch. They are thought to sleep on bent knees, face down, and with hands over their heads.

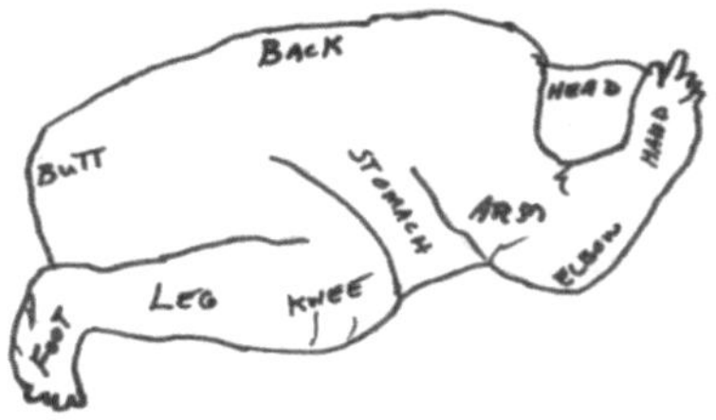

Diagram 8.1: Many believe sasquatches sleep in this position

You might recall an earlier discussion on potential bedding sites found around our campsite from our August 2011 trip. As a point of interest, photos of potential bedding sites are posted in Chapter 4. Each of the possible bedding site photos in Chapter 4 resemble the markings of a sleeping sasquatch as if it had slept as seen in diagram 8.1.

Now then, also somewhat interesting is a large rock found approximately 50 feet behind the root-bed of the downed tree. That rock had embedded quartz within it, suggesting that, under the right circumstances, the large rock could hold a charge or emanate an electromagnetic energy field. This is especially true during a so-called solar storm, or even an electrical storm producing thunder, which, in turn, creates vibrations. I will continue to hint at magnetic qualities and geomagnetic space weather factors throughout the rest of the book.

This mention of quartz and electromagnetism is simply another small piece of a larger puzzle. Also, for your information, Chapter 14 discusses geomagnetism and space weather in greater detail.

October 16: Hiked New Areas Looking for Signs

Nothing memorable occurred during these two days. At least nothing worth writing about, so we'll bump into the next fruitful day.

October 18: Morning — The River

Let me backtrack once more to something that occurred on October 15. As Maggie and I were driving to WF's home, we crossed a wooden bridge over one of the local rivers. Photo 8.5 shows Maggie standing on the bridge looking south along the river.

Photo 8.5: Maggie looking south along a local river

As we drove slowly across the bridge, I spotted a sasquatch walking south along the west bank of the river. Grasses along the river grew tall there, and all I saw was its upper back, shoulders, and the back of its head. It was 150-200 yards south of the bridge; thus, details of how it looked are not forthcoming because of both the distance it was away from us, and the high river grass. I can say, though, that the body hair was a deep brown and it was large. Anything else I might say would be pure speculation. This was my **fifth** view of a sasquatch.

Now back to October 18. Julie and I decided to hike along the west side of the local river where I had seen the sasquatch on October 15. Within 100 yards of entering the area, we found our first sign of a bipedal anomaly in the form of a stick sign; this is seen in photo 8.6.

Photo 8.6: Sticks possibly pointing direction of sasquatch travel

Following the direction of the pointer sticks, we discovered footprints along a trail as shown in photo 8.7, along with additional stick signs.

Photo 8.7: Prints along the trail of a sasquatch

The trail led out of the forest into the marsh along the west bank of the river. We followed sasquatch footprints for another 100 yards and then spotted a couple of possible bedding sites. The possible bedding sites are seen in photos 8.8 and 8.9.

Photo 8.8 is unusual in that it could be of *two* bipedal anomalies sleeping quite close together, which is not something we had seen before. Yet another new piece of data to log and ponder on later. Memorable!

Photo 8.8: Bedding of two sasquatches sleeping together

Photo 8.9 shows a partial bedding site. Obviously, when I snapped this photo, I was not paying close attention to what I was doing. Even so, the photo shows where the head was at the top of the photo, and the shoulders next. This bedding area was fascinating to us because it portrayed a clear impression of a sasquatch. An obvious thought here might be why didn't we cast it? Stated simply, we did not have sufficient materials on hand to do so and the weather was rapidly turning to a cold, fall rain, which would have quickly filled the area with water.

Regardless, evidence of prints, bedding sites, and even scat, convinced me I likely did see a bipedal anomaly on October 15 as Maggie and I were driving across the bridge on our way to WF's home. The signs Julie and I found alongside the river were indeed near the spot I thought

I had seen a sasquatch. In the end, I considered this conclusive evidence I had indeed seen another one. This, then, was the **fifth** living, walking sasquatch I'd seen since the beginning of our sasquatch research in 1999.

Photo 8.9: Partial bedding site; head at top, then shoulder

Photo 8.10: Julie standing in marsh grasses along river

Photo 8.10 shows Julie standing in the tall, brown marsh grasses on the west bank of the river. This photo gives you a sense for why I saw only the head, shoulders and upper back of the bipedal anomaly spotted on our way to WF's home. That is, the grasses along the west side of the river were quite high; for a reference point, Julie is 5 feet 9 inches tall. I suspect that the sasquatch I saw was between 7½ and eight feet tall based on the way I saw its upper back, shoulders, and the back of its head.

October 18: Afternoon — Dimension Lane

On the afternoon of October 18, Julie and I returned to Dimension Lane. We noticed a few sasquatch footprints around Pyramid Rock, so I decided to measure the magnetic energy. Nothing notable. Shucks!

Photo 8.11: Measuring magnetic energy. Photo credit: Julie

Photo 8.12: Julie at Pyramid Rock; she is 5 feet 9 inches tall

October 19: Heavy Rain

Yuk! Way too much water to hike today. Instead, we drove around in the rain checking out areas we had not yet been to; we had an enjoyable time doing this even though it was raining.

October 20: Casts and a Bear Den

This day, Julie, WF, and I hiked yet another location, one I had selected after reviewing my topographic and magnetic maps. The location decision was arbitrary, with me intending to check out a few more of the 1,530,617 acres within the boundaries of the Chequamegon-Nicolet National Forest. About one-quarter mile into the forest, we began seeing footprints. The prints led to an interesting anomaly. That is, we found two footprints side-by-side near some brush, suggesting a bipedal anomaly had stopped at that location to look ahead.

Wondering what it might have been looking at, we carefully scanned the forest ahead of us. We stood where the bipedal anomaly had stood and looked all around. Cool! We saw the entrance to a bear den! We, of course, do not know if the bipedal anomaly saw a bear in the den at the time it stopped to survey the area. We know only that if one had been in the den, the sasquatch would surely have seen it—or otherwise been aware of it. I can say that the sasquatch did not approach the den any closer, because footprints showed that it moved into the forest away from the bear den, heading directly toward a river. Photo 8.13 shows a view of the bear den from where the bipedal anomaly was standing. An arrow points at the den.

Photo 8.13: View of bear den from where sasquatch stood

Photo 8.14 shows a closer view of the den, while photo 8.15 shows the interior. A careful look inside the bear den did not turn up hair or other evidence. Nor did it reveal what may have inhabited it at one time in the past. Apparently, the den had not been used recently.

Photo 8.14: Closer view of bear den

Photo 8.15: Interior of bear den

Photo 8.16 shows me gloved up and prepping to cast the footprints found where the sasquatch had stood while looking at the bear den.

Photo 8.16: Preparing to cast bipedal anomaly prints

Julie and WF hiked further into the forest while I casted the prints. Casting can be time consuming, especially when the ground is wet, and the air is damp; I didn't require help with the casting, so they chose to continue checking out the area.

They followed the footprints to see where they led, but lost track of them after about a mile. I had thought they might find the river as it wound around and through the area. They did not. Obviously, I had miscalculated the distance to the river from the footprint location. Oh, well! Neither the first nor last miscalculation I'll make. They did, however, speak about seeing even more beautiful country, including a provocative rock formation. Sadly, time did not allow the three of us to return to it, so I was not able to explore that particular rock formation. Another day? Another hike? Another year? Maybe.

While waiting for the dental casting material to dry, I wandered about the area of the bear den in a circular pattern. I found nothing further of interest. As soon as Julie and WF returned, we packed the casts, hiked out of the forest to the truck and drove back to WF's home. Another lengthy drive.

This day's hike concluded our October 2012 excursion. Next up, we delve right into the 2013 trips. In total, there were two in 2013: one from late July into the first week of August, and the second in late September. Cool stuff forthcoming!

Chapter 8

[52] Kummer, Chris. "There Was so Much Opposition towards the Erickson Project . . ." Der Wahrheit Auf Der Spur. October 30, 2016. Accessed March 15, 2018. https://www.dietiefe.com/2016/10/30/erickson-project/.

[53] Evidence, Bigfoot. "The Erickson Project Is Now Called The Ketchum Project." Bigfoot Evidence. September 4, 2012. Accessed March 15, 2018. https://bigfootevidence.blogspot.com/2012/09/the-erickson-project-is-now-called.html.

[54] Edwards, Guy. "Sleeping Bigfoot from Erickson Project." Bigfoot News | Bigfoot Lunch Club. Accessed August 07, 2018.

http://www.bigfootlunchclub.com/2011/08/sleeping-bigfoot-from-erickson-project.html.

Chapter 9

A Flag of Youthful Evidence Unfurls — 2013

"The personal life deeply lived always expands into truths beyond itself."

— Anais Nin

We camped in the spot we had discovered on August 21, 2012. It was a bit deeper into the forest and closer to our primary research area. It also seemed more amenable to me for possible visits by sasquatches.

This trip finally solidified my decision to continue my research alone or only with Julie as a partner. You'll soon learn why, and Chapters 12-13 detail dynamic experiences resulting from it.

My internal struggle reached a critical mass point during this trip, prompting me to finally act on the decision to research alone. Of course, certain fireworks arose from this decision; however, I won't bore you with those details. The important point here is that I finally made and acted on a decision long in coming.

July 24: Sounds, Movement, and Cats

Maggie and I had one of our grandsons, Andrew, with us this trip. We spent $2\frac{1}{2}$ days setting up camp; in part, because we had to drive the backroads to find appropriately sized stones to build a proper firepit at this new location. Andrew was extremely helpful with this, for which I was grateful. It wasn't until afternoon of the third day that he and I made it into the forest for our first hike of the trip.

I was hoping he might take an interest in the wilderness and what I do while here. Sadly, this wasn't the case. I suspect he thought I was out of my mind for pursuing such silly ideas as learning about and interacting with a bipedal anomaly. Of course, he may be right; however, I remain unwaveringly diligent in my pursuit of knowledge about the sasquatch phenomenon.

Anyway, shortly after entering the forest with him, I heard sounds ahead of us. I had selected an area not nearly as thick as I often go into while hiking. The sounds, though, came from an area that was a tad thick at the outset. The sounds certainly tickled my curiosity, so off we went into the unknown chasing a silly sound. No wonder Andrew thought I'm a bit batty!

The sounds of something walking preceded us as we moved deeper into the forest. After about 80 or so yards, we began seeing prints more clearly and they surely appeared fresh. This, of course, really excited me, for here was further evidence of a bipedal anomaly not more than two miles as a crow flies from our new campsite. So cool! And Andrew was with me! Sweet!

We followed the prints through the forest and onto an overgrown, incredibly old logging road that appeared almost as a narrow meadow. The prints measured 20 inches long, $6\frac{3}{4}$ inches at the heel, and $5\frac{1}{8}$ inches at midfoot. The stride measured 5 feet 3 inches, and it appeared as if the bipedal anomaly was walking slowly, without fear or concern for who or what was behind it. For some unknown reason, I did not measure the step. Oh, well! Anyway, the bipedal anomaly kept moving in a northeast direction, which was generally toward a local river.

We continued to hear movement noises as we pushed forward. The narrow meadow had a slight incline on its north side. As we walked, I noticed a fresh cat track that measured three inches long and four inches wide. It looked as if a large cat had crossed behind the bipedal anomaly and it seemed to be moving rapidly out of the area. It apparently wanted to get the heck out of there. Hmm! Did it know something we didn't?

This was Andrew's first exposure to deep wilderness. Imagine the look on his face once he realized we were not only following an enormous sasquatch with a huge footprint but were also not far from a decent-sized cat. Now then, I doubt the cat would have bothered us; it seemed in too much of a hurry to leave in a 90-degree direction away from the bipedal anomaly. Nonetheless, it seemed prudent at the time to end our hike and return to the truck.

I handed Andrew the topographic map and suggested he look at it, figure out our current location from the surrounding terrain by comparing it to the grades and elevations presented on the map, and then lead us out of the forest in as straight a line as possible to the dirt road along which we had parked the truck. I was immensely proud at how well he read the topo map! This was his first attempt at it, and he was very successful. Yay!

While hiking out, he stopped and snapped a photo of what I believe could be a purple star thistle entertaining a monarch butterfly. I may be wrong on both counts; even so, the photo turned out well and is posted as photo 9.1. I was deeply touched that he showed a natural affinity for enjoying the normal beauty of nature abounding within the Chequamegon-Nicolet National Forest—even knowing that a sizeable cat and a large sasquatch were nearby. Very proud grandpa.

Alas! This hike seemed to be the catalyst for Andrew's decision to not go out again. Although disappointed, I honor his courage for being able to discuss his wishes and needs with me while explaining his reasoning. Clearly, tramping through thick wilderness looking for something that many swear cannot exist is not for everyone.

After our first hike, he and grandma had a wonderful time doing other things while I bounced rather aimlessly around in the forest, almost like a ball in an old pinball machine with a wildly active person at the control buttons. Goodness! I sure was anxiously awaiting Julie's arrival on July 28.

Photo 9.1: Butterfly on purple thistle — Photo credit: AJS

Oh, yes. One more item to mention here. Because Andrew was clearly not captivated with bipedal anomalies or large cats, I chose to not tell him about the pointer stick marker that was only about 20 feet from the back of his tent! It pointed *directly* at his tent, and had he not moved into the trailer with grandma and me, I'm convinced he would have had an overnight visit from a bipedal anomaly. Maybe even a subtle touch of one through the tent wall! Hope he will forgive me when he reads this. At the time, though, I felt it best to keep this knowledge to myself, although I later shared it with Maggie ... much later and, no, she was not happy about it.

Guests showed up from July 25-27; hence, no research was conducted. Very disappointing! Especially considering I had a finite amount of time in this northern area to engage in research. Again, way too many distractions for a serious research project.

That was it for me. I told myself that changes were forthcoming! Guaranteed!

July 28: Julie arrived, and we returned to Dimension Lane

Julie and I drove to Dimension Lane on July 28 and noticed a couple of obvious changes. First, fresh footprints were seen along the length of Dimension Lane and at Pyramid Rock. Clearly, a bipedal anomaly had stood at Pyramid Rock for a time. More interesting, though, was the movement and replacement of five stones around the bottom of Pyramid Rock.

I have dozens of photos of the Dimension Lane area taken over the past couple of years, so it was a simple matter to figure out which stones had been moved and their new placement. Plus, we easily detected the areas from whence the five stones originated due to indentations left in the soil after they were moved.

The five moved stones were placed at the north-northeast side of SSE13, which, as you might recall from Chapter 7, is Pyramid Rock. They were set among a few low-growing ferns. Photo 9.2 shows the new placement of the stones.

Photo 9.2: Stones placed a few feet away from Pyramid Rock

Although I do not "read" bipedal anomaly, the placement of the stones looked vaguely familiar. That is, they slightly resembled the cat track my grandson and I had seen four days earlier! While it was not unusual to see sasquatch signs at Pyramid Rock—or anywhere along Dimension Lane for that matter—an intentional movement and replacement of a series of stones fell directly into the category of "Excuse me?"

If parts of Dimension Lane were used by sasquatches for communication purposes, it's entirely possible that the pattern in which these stones were placed was meant to warn other sasquatches of a cat prowling the forest. Although I don't know this for sure, it seems reasonable that sasquatches have and employ a *warning system* of some sort to advise others of potential danger. Photo 9.4 shows the path that the bipedal anomaly took when it left Pyramid Rock to wander back into the forest. The movement and replacement of stones leads to another proposal:

Proposal 5:

> Sasquatches use rocks, stones, sticks, etc., as forms of communication.

Photo 9.3: Path bipedal anomaly took leaving Pyramid Rock

Another unexpected incident occurred today, albeit back at camp. As background information, we take an old logging road into the forested campsite area. Across the logging road from camp is an area of thick shrubbery. I walk across the road and hide behind the thick shrubbery to pee.

I walked across the road after dinner and my goodness! Right where I usually stand to do my thing is a set of large footprints. I'd post photos of them, but I don't want to give away my outdoor urinal location! Regardless, I'm not sure if a bipedal anomaly took a leak there, but the footprints certainly showed where it had stood *directly* over the top of my prints. Hmm! Territorial? No worries; I will simply find a different outdoor urinal! In this instance, BIG wins!

This day concluded the July 2013 portion of the trip. Maggie and our grandson headed home while Julie and I stayed at camp to pursue continued research.

Before she left, I shared my thoughts with Maggie about no longer having others with me during my research time. At first, she was less than accepting of the idea, naming a long list of potential things that could possibly go wrong if I were alone. Clearly, she had put thought into it since I had brought the idea up many times over the past few years. I understood her concerns; however, I can be somewhat obdurate at times, and although I knew it was past time to create and honor clear boundaries around my research, I truly did **not** want to offend anyone.

Regardless, I strongly believed that if future research was to produce effective results, I needed to be alone when Julie wasn't available to research with me. The discussion with Maggie was quickly turning into a heated debate when we finally tabled it for a later date. I struggled with this internal battle for a while longer before ultimately creating clear and strongly held boundaries for future research efforts. You will see my internal struggle play out, and eventually conclude as we continue this story.

August 2013 — A Time of Clarity

Julie's and my time together during early August 2013 turned out to be singularly useful and pertinent to the ongoing realization I needed to work alone or only with her. For instance, Julie and I have an unusual synergetic relationship, including the fact we do not always have to speak to know what the other is thinking. This can be of enormous value while traipsing through the wilderness and stepping regularly into unknown situations.

August 1: Reconnaissance kayak trip on a local river

Photo 9.4: Approach to Baby Gate

Julie and I headed out to kayak a local river south on a reconnaissance outing. Along the way, we discovered an interesting area we dubbed "Baby Gate." You'll soon know why we chose this name. Photo 9.4 shows our first approach to the Baby Gate area, while photo 9.5 is a shot of the entrance off the river.

Photo 9.5: Baby Gate Entrance

Photo 9.6 shows footprints of *two* bipedal anomalies, with one overlaying the other. If you look closely at the prints, you will see what looks like a crippled adult sasquatch foot print overstepped with what I believe is a young sasquatch footprint.

The toes of the adult are obvious and appear to be shaped similar to a human foot. It looks to me as if the imprint came from the adult sasquatch's left foot; I base this thought on the placement of the big toe.

The overprint of a young sasquatch looks even more like a human foot, and it is clearly more symmetrical than the crippled adult foot.

Photos 9.6 and 9.7 are truly amazing photos, and we consider ourselves fortunate to have captured them. We were on opposite sides of the river and Julie's eagle eye spotted them. She called me over for a looksee, and boy did we looky-loo at these suckers!

Photo 9.6: Prints near entrance to Baby Gate

We were thrilled to capture these dual prints in a photo. Unfortunately, they were too wet to cast. Even so, the photo is quite spectacular, eh? You can soon view all the color photos posted in this book on my website, sasquatchandme.com.

We searched the surrounding area for castable footprints; however, the prints we found close to the river were all filled with water, and those discovered farther inland had too much high grass bent over, making them unsuitable for casting. We tried draining some of the footprints near the river. Sadly, water kept flowing back into them. Bummer!

Photo 9.7—as noteworthy as photo 9.6—shows several footprints of the young one walking toward the river where the wooden sticks were driven into the bed of the river. We believe the sticks served as markers, probably to inform the young one to stay out of the river. Truly, another remarkable photo. Photo 9.7 also appears on the front cover of the book.

Photo 9.7: Prints of young sasquatch walking towards river

In the end, we considered this reconnaissance river trip well worth our effort and time. At the date of this writing (June 2018), we remain elated to have captured these prints in photos.

August 02: Kayaking south on a local river again

This day, Julie and I headed farther south on the river, paddling well past the Baby Gate area toward a location we'd selected from my well-used topographic map. We paddled for about 1½ hours, portaging several beaver dams along the way. The river flowed through a sizable swamp near the area we wanted to explore, so landing where we'd be able to walk on solid terrain was challenging. Nonetheless, we persevered.

As we approached our desired spot, we wound our way through the swamp, testing possible stopping areas for solid ground. Eventually, we made progress only by pulling on the swamp grass to move our kayaks through the marsh until we were able to get out and walk on semi-solid land. Squishy, but doable. Thankfully, we were wearing muck boots; photo 9.8 shows the debarking point.

Photo 9.8: Landing area. Credit: Julie

We hiked up the small wooded hill seen in photo 9.8 and were pleasantly surprised. It was a delightful, serene area, with a nearly sacred quality to it. The area was complete with a thought-provoking arrangement of large rocks. Once again, the rocks held goodly amounts of lovely, sparkling quartz, hinting, of course, at a potential for magnetic anomalies to occur. They did.

The next three photos show large rocks with quartz in them. These rocks did not look as if they were arranged in a specific order; rather, they appeared randomly around the hilltop. Each rock was large, prominently displayed, and readily accessible to all who came to this charming spot. The most interesting rocks we found are discussed after the following photos.

Photo 9.9: Heart shape with quartz embedded in large rock

Photo 9.10: Large rock with embedded quartz

There was one arrangement of rocks that seemed a bit unusual. This arrangement was oriented in a way that suggested both clear intent and strong use of intelligent reason with respect to their placement. I'll not purport bipedal anomalies placed the rocks in the fashion we found them; however, I am convinced that *something* or *someone* did so at some point.

Photo 9.11 shows Julie sitting on a flat stone lying directly between two large rocks amidst four pine trees. She is facing south.

Photo 9.11: Julie sitting within an unusual placement of rocks

Photo 9.12 is a shot from the back area of the rock placements. Julie's arms are spread apart to show distance with the rocks that she leaned against. Again, this was quite a unique placement of the rocks! Looking closely, you should see how the rocks appear atop a build-up of soil between the four pines.

I don't know how to adequately interpret this, other than to remark that Thom Powell, in his novel *Shady Neighbors*, (2011), proposed that he and a neighbor once found a spot where stones were placed in a manner suggesting a gravesite. I doubt, however, that a gravesite lies beneath these rocks. Other large rocks around the area, though, could easily be interpreted as possible gravesites. Regardless, a fascinating place for sure!

Moreover, I do not know how bipedal anomalies dispose of their dead. From their size, though, I'll put forth quite readily that they could certainly dig a grave to bury their dead and then place huge rocks over a gravesite to protect it. Merely a considered thought; nothing more.

Photo 9.12: Backside of unusual rock placement

Magnetometer readings in the area showed readings well below Earth's typical range of magnetic values. Again, I'm not sure how to interpret those readings. That is, other than to point out that the value measured on the flat stone between the two large rocks measured a goodly bit below the typical value of 50,000 nT (500mG). For instance, Photo 9.13 shows a reading from this area; a milligauss meter showed

311.7 mG (31,170 nT). Wow! A low enough value to cause a near automatic experience with coherence! Sadly, not having the equipment to measure coherence via HRV biofeedback left me without the ability to prove it.

At any rate, it's no wonder that the entire area felt serene and nearly sacred. In fact, Julie and I felt in complete harmony with all around us the entire time we were there. Recall that earlier heart-head research repeatedly proved that readings this low automatically snapped an individual into physical heart-brain coherence, complete with a subjective feeling defined as spiritual. We certainly spoke quietly the entire time we were there, almost as if we were in a temple. Hmm! Maybe we were in one of nature's myriad outdoor temples where sasquatches engage in rituals sacred to them? Who knows? Not us.

Photo 9.13: Milligauss meter reading at stone seat area

This day effectively concluded Julie's and my time for the August research. We paddled leisurely back to the boat launch, loaded the truck, and returned to camp. Julie headed home late that day, and I broke camp the following morning. The next trip north occurred in late September 2013.

September 2013 — Not Really a Research Trip

Maggie and I, along with Brushtracker and Dragonfly, arrived in the research area on September 26, 2013 to check out and evaluate another section of the Chequamegon-Nicolet National Forest within the Northern Highland geographical area. We rented two rooms at a rustic motel named Little Pine Motel in the small village of Hiles, Wisc. Little Pine Motel is a picturesque, pleasant, 12-room motel in the middle of the Chequamegon-Nicolet National Forest. Easily seen purple flag-type signs hang on power poles and flutter in the breeze as one drives through the village. The flags proclaim Hiles, Wisc., as the "Heart of the Nicolet." A neat concept, eh? A couple named Mark and Julia own and run the motel. They are positively delightful and had no clue what we did while camping in the wilderness or the few times we rented rooms from them. Oops! At least they had no clue until now!

From Little Pine Motel, I was able to branch out into the vast reaches of the Northern Wisconsin geologic area with ease. Everywhere we research is a long way from everywhere else; hence, a central jumping off place was needed, and Little Pine Motel filled that need. Sweet!

Mark and Julia cater to the fishing crowd, four-season hunters, and, in the winter, snowmobilers. And, of course, to weirdos like us. They are indeed a charming couple and treat their guests with high respect intermixed with an indigenous local allure. If you are ever in the vicinity, it's well worth a stop, even if only to visit with Mark and Julia. If you ever do stop by there, tell them Maggie and Roger sent you. Hopefully, they won't boot you out!

September 26-27: Aimless Wandering

On September 26-27, the four of us wandered rather aimlessly around the surrounding area, acting more like tourists than researchers. We drove far enough into the forest to check out Dimension Lane and found it to be the same as it had been in August, including the five stones that looked like a cat track. They thought this was a stimulating anomaly.

September 27: Brushtracker, Dragonfly, Maggie, and I on a local river

This day, Brushtracker and I were in kayaks, while Maggie and Dragonfly were in a two-person rented canoe. Little did I know at the time of departure that the women were not the canoeists they claimed to be. It didn't take long, though, to discern their unskilled attempts at canoeing. We headed to the same area Julie and I had discovered on August 2. We moved slowly along the river due to the lack of paddling experience shown by Maggie and Dragonfly.

This wasn't all bad, though, because they were able to see and comment on bipedal watering spots, areas we would likely have missed had we been paddling rapidly. In fact, we saw rather clear prints at one spot, which always starts hearts to beating quickly. Photo 9.14 shows the prints, and photo 9.15 points out the path a bipedal anomaly took as it left the river.

Photo 9.14: Prints at the edge of the river

Photo 9.15: Path bipedal anomaly took away from the river

About half-way to our desired location, we stopped for a brief rest and to consume light snacks. A pleasant way to spend time with friends, especially if the goal is fun, not research. Following a brief rest, we returned to the river and paddled quietly to our destination.

Having arrived at our destination, the four of us enjoyed the beauty, serenity, and seeming magical qualities of the oddly arranged stones. The sun was playing hide and seek with the clouds, which caused the quartz in the rocks to intermittently twinkle, creating a subtle feeling the rocks might very well be winking at us. Quite the welcome, wouldn't you say? Don't know about the others, however, I playfully winked back. Gosh! Hope this doesn't mean I was flirting with the twinkling quartz! Maybe I'd spent too much time in the wilderness? Hmm!

We took turns sitting on the stone seat between the two large rocks seen in photos 9.11 and 9.12, each enjoying the privacy of the moment within the confines of quiet thoughts. Quaint? Perhaps. A private moment or two with our picturesque, secret thoughts? Indeed! Really a very lovely spot, and perfect for one to temporarily lose oneself as the gentle sweep of nature's soothing beauty peacefully washes over you, thereby granting a feeling of being enveloped in the arms of forest angels.

I took no measurements this trip. Rather, we were here simply to share the scenic beauty and serene feeling of the area with Brushtracker and Dragonfly. They claimed to have experienced the serenity of the place. I trust they enjoy memories of the area and the trip as much as Maggie and I do. Memories, it turns out, not only spark fond feelings for friends, I believe that they serve to spur one on to greater future experiences.

After enjoying the wonder of this scenic area, we knew we had to head back to the boat landing and the truck. Oh, my! We had dallied along the way here. Now it was time to paddle hard if we wanted to be off the river by dark.

So away we went—slowly, and for Dragonfly and Maggie—clearly in a cumbersome manner. They were tired; hence, they were even less able to paddle than they were on our way in. Oh, my. They began bouncing riotously from bank-to-bank as they seemed to fight each other's attempts at paddling. Back before dark? Goodness! Seemed like extremely high hopes to me. How about sleeping on the river? We had personal flotation devices (PFDs), but where were the sleeping bags?

Maggie and Dragonfly were deplorably inexperienced canoeists. Following a few more painful minutes of watching their pitiful paddling, I realized they needed to be towed. Fortunately, I always carry a length or two of rope with me in my water conveyance, so—I hooked them up to my kayak—and tow them I did. This was a challenge, yet I really wanted to be off the river by dark.

We made the boat landing just at dusk, debarked, loaded the water conveyances, and headed back to the motel. The three of them went out to dinner while I stayed behind to rest. The river trip was a great time with friends; however, it surely taxed this ole' boy to the max on the paddle back.

This concluded our September 2013 jaunt into the wilderness, and we headed home the following day. Another trip under our belts! This trip, however, was spent with friends—not research—so my ongoing issue of researching alone did not interfere with the good time had by all.

Chapter 10

Shining a Light on *Habituation* and *Familiarization* — 2014

"Success is not final, failure is not fatal: it is the courage to continue that counts."

— Winston Churchill

The year 2014 was physically a bit challenging. I had heart issues again, so was unable to do as much as I wanted. Even so, this year shined a spotlight on something known within the sasquatch community as habituation and familiarization.

Who knows? Perhaps the heart issue that kept me close to camp much of the time during the 2014 excursion was a blessing in disguise. This, because staying around camp taught me that bipedal anomalies—for whatever reason—*found me* rather than me going out to look for them. Oh, for goodness sakes! All these years of thousands of miles of driving and hiking, and I discover those suckers would come to me? What next?

Lesson learned and appreciated; well, appreciated. It sometimes takes a bit or so before I fully learn and absorb lessons. Anyway, this year certainly gave me a clear introduction to habituation and familiarization. By 2017, however, habituation and familiarization **ruled** the research.

Fascinating and dynamically thrilling stuff to come. I can hardly wait until we reach the point of sharing future tales.

September 2014 — Special time with Maggie

September 05: Setup camp

Even though Maggie does not research with me the way she once did, she will always spend a few days with me in the wilderness. Too, although my research time while Maggie is with me is limited, she occasionally enjoys getting out with me. She and I setup camp and wandered the woods around the campsite checking for signs, either old or fresh. We found both, so settled in for the night with a sense of high expectation for the forthcoming days of this trip.

September 06: Scouted the area

This day, we drove all around this year's research area, stopping from time to time to check for bipedal anomaly signs in spots known for repeated visits by more than one sasquatch. Per usual, we saw fresh signs at each of the stops. This, of course, bolstered my spirits; I always feel excited when I see fresh signs of bipedal anomalies.

One of our many stops was a boat landing on a local river. Yikes! The river was quite high this fall, and I wasn't sure if I'd be able to kayak under the bridge to reach an intended area of research. Time would tell.

The Chequamegon-Nicolet National Forest area had received a good bit of rain this year, causing rivers, creeks, and streams to rise higher than they had during the past few years, in places rising completely out of their banks. Again, it's too soon to tell if I can paddle a kayak under the bridge to an area I really wanted to check out.

We returned to camp in early evening, prepared dinner over the campfire, and settled in for a quiet twilight. We were in the same camp area as earlier trips; however, I had parked the trailer even deeper into the woods. The forest now formed a half-circle around the trailer and the open space between the front of the trailer and the woods averaged 34 feet. The backside of the trailer was only about five feet from the edge of the forest, with just enough room to walk behind it so I could readily add water to the tank, etc. That's right; I am now truly in the wilderness.

Coals in the campfire were sizzling, creating a false sense of circular light exploring tentatively upward into the onset of dimness and, at times, spitting gases out of the remaining small wooden knots. Sizzle and spit … sizzle and spit, a tiny dance of flames bidding us a quiet goodnight. A comforting feeling indeed; and this, while enjoying the exquisite beauty of nature's gentle tread into the oncoming darkness of night. It's hard to beat an evening like this while sitting quietly next to the love of your life.

By now, the sun had set, painting the open sky with orange and reddish tints—with a smidgen of lavender thrown in for variety—all of which invited quiet oohs and aahs from Maggie and me. While the open area in front of the trailer had a bit of light left over from the sunset, the surrounding woods were dark. Not gloomy; not threatening, simply dark.

Of an immediate sudden, the gentle, typical crooning sounds of night stopped. No spattering of a leisurely retreat of sounds. All nighttime forest sounds simply stopped. All was instantly quiet. Maggie and I looked at each other in wonder. Did this mean what I thought it meant? Or was my imagination rapidly wandering into a spate of delusional nonsense?

Then we heard it. Movement in the surrounding forest. It was a deliberate sound. Not the slight whispering of something quietly moving through the forest with intentional stealth. No, the sound was deliberate. Maggie grabbed my hand, gasping nearly inaudibly while staring at the forest. What was it? But wait! Another sound tapped lightly on our eardrums, this time from a different direction. Maggie's grip on my hand tightened … and then there were three!

What in the world? By now the forest was too dark to see into at all; we could only stare intently into a dark void, wondering who or what was out there. With swiftly beating hearts, we gaped keenly into the deepening darkness, trying our best to fathom a shape, or anything for that matter. Instead, we heard movement from three distinct directions around our half-circle, which no longer held much light. The campfire continued to sizzle and spit, periodically shooting microscopic bits of tiny hot light toward the night sky, almost as if casting out a welcome to the sounds of deliberate movement coming from the ethereal darkness of the forest.

After several minutes of listening to the sounds of physical movement penetrating our now very tiny half-circle, we heard a soft grunt. This came from the one which arrived first. Not knowing what this sound meant, we continued to listen and stare intently, listen and stare intently, listen and…

We then heard what I can only describe as a chortle. This came from the third one to enter our area. Nothing from the second one; only the first and third. Maggie, still gripping my hand as if it were a lifeline to a so-called non-delusional world, looked at me and said, "Think I'll go inside." Okay. As we stood up from our chairs, the sounds of movement ceased. Once she was comfortably inside, though, I quickly returned to sit in the same position I had occupied earlier.

I heard no movement sounds; nothing. The purity of silence reigned. Movement sounds did, though, resume after a few minutes of deathly still quiet. Clearly, whatever was out there wanted me to know they were there because, while they moved somewhat stealthily, they seemed to make a bit of noise by design. Okay. I know three of them are there; now what?

I chose to not shine a light into the forest because I wanted whatever was there to believe I meant no harm. Besides, seeing what was making those ghostly sounds might have scared the scat out of me!

Whatever was there continued to move about just outside the camp area. I certainly did not feel a sense of hostility coming from whatever was there; rather, simply an inner knowing that they wanted me to be aware of them. No, I do not know why they wanted me aware of them, only that they did. I was undeniably aware—and felt exhilarated because of it. The slow wandering movements in the forest finally ceased after about an hour. Shortly after that, typical night sounds began again.

Way cool! Time to turn in, sleep, and await the coming of the following day's events, whatever they might be.

September 07: Clear evidence of visitors

This day found me up early. I went into the forest, walking a large half-circle outside the open space of camp. Sure enough! I discovered footprints. This, of course, confirmed we had been visited last night. At least the evidence convinced me; how thrilling.

A careful perusal of the area produced three distinct sets of prints. At this point I decided it was likely that the three bipedal anomalies from the night before belonged to an adult male, an adult female, and a young one, probably a single family of sasquatches checking us out as they wandered the forest. Maybe the parents wanted the young one to see the goofs camped in the middle of the wilderness? Who knows?

Before I go further with this finding, let me state unequivocally I have no clue why they showed up the previous night. No clue what caused them to come to this area at the time they did. No clue from whence they came, only their general direction. The prints, though, surely confirmed their visit to our humble camp in the depths of the wilderness where we chose to set up our base.

I later followed the footprints to figure out the direction they took when they left. The young one was the hardest to track; maybe because it didn't weigh as much as the adults and left less of an impression on hard ground. Anyway, I decided they came into camp from the north-northeast and then left in the same direction after spending of bit of time checking out us humans and, from time to time, chortling at us. This led me to scrutinizing my many topographic maps.

Eventually, I decided they were likely headed toward a local river, which, as the crow flies, was approximately five miles from camp. By road, the river was about 20 miles away. There's a lot of thick forest and swamps between camp and the river; probably not a big deal to the bipedal anomalies, but for me to track them that far through that terrain would take a serious amount of time as well as physical effort. Hmm! Not today. Not saying I was lazy; merely that I didn't feel like tackling a strenuous and lengthy hike.

I measured the three sets of footprints. The adult male prints measured $21\frac{3}{8}$ inches long, $6\frac{1}{8}$ inches at the heel, and $5\frac{3}{4}$ inches at the midfoot. The adult female measured $19\frac{3}{4}$ inches long, $5\frac{7}{8}$ inches at the heel, and $5\frac{1}{4}$ inches at the midfoot. The young one measured $13\frac{1}{2}$ inches long, $4\frac{7}{8}$ inches at the heel, and $4\frac{1}{4}$ inches at the midfoot. Stride and step were not measurable; also, the footprints were not castable. I followed their tracks for about 80 yards, then returned to camp. I swear, regardless of where we hike as we explore the Chequamegon-Nicolet National Forest, the spectacular beauty is not only exquisite, it wraps its ever-reaching tentacles of unfathomable, unseen energy around your heart, guaranteeing that you must return for more.

That was it for research today. We spent the afternoon driving around the large research area again, which can take a few hours if driving slowly. Per usual, we stopped several times to look for signs of bipedal anomalies and found quite a few. Always exciting, always encouraging!

We spent that evening much as we had the night before; that is, dinner cooked over the campfire and then sitting quietly outside during the evening. Disappointingly to me, no visitors made themselves known that night, so we called it quits around 10:00 p.m. and went to bed.

Around 2:00 a.m., we awoke to a gentle rocking of the trailer, almost like one would rock a cradle with a baby in it. Maggie became slightly concerned; however, the gentle rocking felt soothing to me.

The rocking lasted five minutes or so, and although I felt completely comfortable with it, Maggie remained a tad anxious wondering if they were going to tip the trailer over or damage it in some other way. After a bit more time and a reassuring hug or two, we drifted back into the dreamland of sleep, knowing that the gentle rocking was real, and not part of a collective or delusional dream. After all, Maggie and I had communicated while awake … hadn't we?

September 08: Prints near the trailer

Goodness! We found footprints right near the front of the trailer this morning. Okay. So, the rocking *was* real! What to think? What to do? Again, the rocking during the night was clearly gentle, suggesting nothing to worry about. That is, it certainly didn't *feel* hostile, angry, or intimidating at the time.

We had heard nothing during the rocking. No vocalizations … no sounds of any type. Simply a gentle rocking. I thought it was quite an amazing experience and after long discussion, we agreed the sasquatch responsible for rocking the trailer merely wanted us to know it had returned and was not troubled by our presence. This was my interpretation. I'm clueless as to the sasquatches' interpretation and chose to not ask Maggie to elaborate on her thoughts.

Photo 10.1 shows the prints we found at the front corner of the trailer. The heels were placed right at the front corner, so it seemed to me that the sasquatch used its backside to rock the trailer while leaning against it. These prints measured the same as did the adult male from the evening of September 6. That is, they were 21⅜ inches long, 6⅛ inches at the heel, and 5¾ inches at the midfoot.

The remainder of this day was spent visiting friends and taking Maggie shopping. Regardless of where one drives in a four-county area, the scenery is both beautiful and inspiring!

Photo 10.1: Prints at front corner of trailer after the rocking

September 09-10: Additional prints near the trailer

On the morning of September 9, Maggie stepped out of the trailer and was at once mesmerized with a monarch butterfly angelically wafting around her. She was thrilled when it landed on her, especially when it moved from her jeans to her hand; see photos 10.2-10.3.

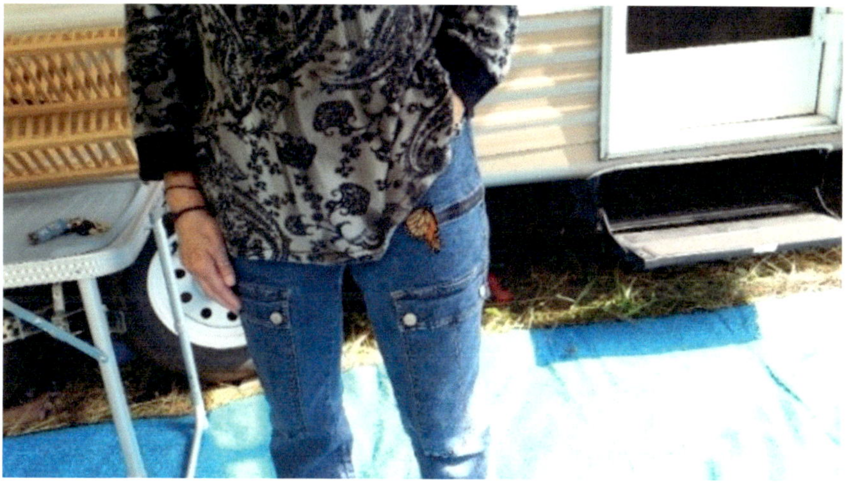

Photo 10.2: Monarch butterfly landed on Maggie's jeans

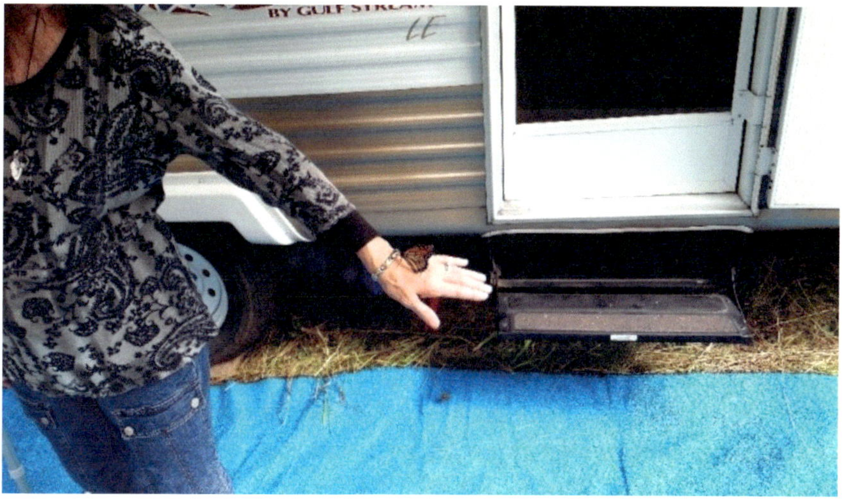

Photo 10.3: Monarch butterfly moved to Maggie's hand

An interaction like this with a wonder of nature—minor though it might seem—reflects a bit of the joy that occurs when one takes the necessary time to consciously appreciate the mild side of nature's ever inviting, interactive beauty. Gosh, but I love our times spent in the forest!

Alas! The busyness of daily life, the persistent stress of living in an ever-evolving, high-tech society, forever striving to keep up with or on top of the seeming daily deluge of new and better or faster tech devices, texting, getting lost in Facebook, Twitter, Instagram, YouTube, Snapchat, or myriad other apps, far too often distracts us from taking time to quietly connect with the tender, healing embrace of nature. Too regularly we forget to momentarily set aside our myriad high-tech devices and succumb willingly to the fresh breath of life that nature so willingly provides. To me, a momentary experience with nature's breath reminds me that the life I live is unquestionably driven by the choices I make.

Choosing to include a bit more time with nature can be naught other than beneficial when nature's subtle, meaningful interactions teach us to appreciate not only the fast pace of daily life, but the blessing of a quiet moment with nature's embrace. Taking time to be still with nature should not be hard to do; rather, it ought to be as typical to us as grabbing a smart phone to text, to Facebook, to tweet, to Instagram, to YouTube, to Snapchat, and so forth. If it's not, consider taking the time for a nature break. You might be pleasantly surprised with the results, and your smartphone can take irreplaceable photos creating memories for later years down the highway of life.

This day, we drove around other research areas. The Northern Highland geographical area of northern Wisconsin is such beautiful country, with more than 1.5 million acres of forests worthy of checking out, including little-used roads, old logging roads, and overgrown trails, not to mention the vast areas where few people ever venture because the hiking or paddling is quite strenuous.

We stopped at a couple of places, hiked for a bit, saw a bear cross the road in front of us, and I'm here to say that sucker was in a hurry! I doubt it even knew we were on the road with it. Man, oh man. Those bears sure can travel once they kick it into warp speed! Seeing this bear run accentuated the idea that one can **never** outrun a bear once it kicks into warp speed! I mean to say that bear was booking so fast a single blink would have caused us to miss it! We also saw two bald eagles and a hawk. Gosh! I sure do love these wilds.

September 10 was much the same, although we took a couple of longer hikes this day. One hike was in an area that Julie and I had checked out the year before. We found many footprints, saw sticks in the form of markers, several unusual stone markings, and an obvious sasquatch trail that seemed quite well-used.

I suggest well-used because this trail had matted down grasses from so many prints it was impossible to distinguish individual ones, and the trail was cleared at a height in keeping with the height of bipedal anomalies. See? I did learn to look up! This trail was about 3½ miles from camp as the crow flies. We stopped at a couple of large rocks for snacks and water.

A real benefit from this hike included Maggie watching me follow clear and necessary woods safety procedures. I feel this left her with the impression that when I'm out alone I do indeed honor caution and safety rules of the woods. I also taught her how to use a compass for bearings in an unknown, thickly forested area.

She surprised herself with how well she did once she started getting the hang of it. Too, this short lesson in compass reading augmented her flowering belief that I can be safe while alone in the wilderness. Anyway, photo 10.4 shows Maggie pretending to rest, or was she simply posing? Either way, her delightful smile shows she enjoyed this hike and our time together.

Photo 10.4: Maggie resting on large rock during a break

We returned to camp, cooked dinner over the campfire once again, and then sat quietly—long into the evening. We heard visitors; this time though, only two. They did not come as close to camp as before, yet they were still out there informing us of their presence through intentional movement noises. We went to bed at 11:50 p.m.

September 11: Shopping and sightseeing in local villages

Today we went shopping, sightseeing, and drove through several local villages. That's it; except for calling Julie and wishing her a happy birthday. Maggie greatly enjoys browsing through thrift stores, and northern Wisconsin has enough of them to quell her need for a periodic fix. She enjoys this so much that I often take her "thrifting" just to see the expression on her face when she holds something up that I might think is junk and she claims it as a lasting treasure. Of course, I think seeing and hearing sasquatches is a lasting treasure, while she is not quite as enthusiastic about this as I. Balance? Who knows?

This day concluded her time with me for this trip. She flew home to handle personal business, such as visiting with our children and grandchildren, and tending to our cats and the wild birds and squirrels she feeds.

September 12: Yuk

I didn't feel well today, so hung out at camp. Nothing to report.

September 14: Kayaking

I went kayaking today, paddling a local river and then a creek branching off it. The river was indeed high enough that I had to duck going under the bridge near the boat landing. Glad I didn't tilt the kayak too much and dump myself and gear into the river. The water was not warm.

Julie and I have spent a goodly amount of time on local rivers and creeks in our research area. Because of this, it seemed important to inform ourselves about the flora surrounding and encroaching onto the banks of the rivers and creeks. My personal take on this idea stems from the thought that the more I learn about a research area, the better equipped I am to understand the signs, movement, and behavior of bipedal anomalies. Therefore, when probing an area for research, I study the flora, fauna, geology, hydrology, history, and even anthropological finds if any are available.

The flora closest to the rivers and creeks in the Northern Highland geographical area are mostly a combination of grasses, sedges, and rushes. I am not a botanist, so will not embarrass myself pretending to distinguish between the myriad varieties of grasses, sedges, and rushes. Even so, the WI DNR informs us that there are 12 common sedges found in that section of northern Wisconsin. Furthermore, river marshes in the area commonly produce narrow leaf cattails, bulrushes, pickerelweed, lake sedges, and/or giant bur reed. The most common rush in this area is the so-called *bog rush*.

Moving away from the near riverside flora onto drier land, we are introduced to an abundance of many trees. For instance, once again, according to the WI DNR:

> The shorelines [of the local waterways] are heavily wooded with sugar maple, yellow birch, hemlock and white cedar on the upper stretches and aspen, silver and red maple, white and jack pine the most common trees on the lower reaches.[55]

The WI DNR also informs us that the local rivers "are born in the vast forests and swamps of the Chequamegon-Nicolet National Forest."[56] I submit for consideration that one solid reason the area in which we research is rife with signs of wildlife and bipedal anomalies relates to the following quote, once again, attributed to the WI DNR:

> The entire … length of the [local rivers and] major [tributaries] ... were designated by the Wisconsin legislature as State Wild Rivers … to be protected from development and kept in a natural, free-flowing condition.[57]

Without a doubt, protected rivers, creeks, and streams produce sustainable wildlife more readily than those areas not protected. Although I'll not go into depth on the subject here, I also propose that knowing which amongst the local flora offers medicinal value in the areas you wander while seeking bipedal anomaly signs can be helpful to planning excursions into the wilderness. This, because I believe most wildlife and assuredly bipedal anomalies instinctively gravitate to medicinal plants to aid in their longterm survival.

Now that I've named some of the flora within our area of research, let me briefly address the idea of changing magnetism's possible alterations to plants. Much of this book is about the way magnetism and geomagnetism affect sasquatch behavior. Interestingly to me, scientific evidence suggests that Earth's geomagnetic field also produces effects on plants. For instance, Massimo Maffei, a plant biologist at Università degli Studi di Torino Turin, Italy, wrote:

> The geomagnetic field (GMF) is a natural component of our environment. Plants, which are known to sense different wavelengths of light, respond to gravity, react to touch and electrical signaling, cannot escape the effect of GMF...[58]

Although fascinating by itself, Maffei's next statement lends further credence to the idea of magnetism's effect on plants:

> Revealing the relationships between MF [magnetic field] and plant responses is becoming more and more important as new evidence reveals the ability of plants to perceive and respond quickly to varying MF by altering their gene expression and phenotype. [Say what?] The recent implications of MF reversal with plant evolution opens new horizons not only in plant science but also to the whole biosphere, from the simplest organisms to human beings.[59]

Whoa! Obviously, even flora within an area of research is susceptible to changing magnetism. Goodness! I truly hope that the further we progress into this story, the more aware you become about the importance of magnetism, geomagnetism, solar events, etc., to my research of sasquatches.

Okay. Enough digression for now. Instead, let me post a couple of photos showing a bit of the pristine riparian flora along one of the creeks I kayaked this day.

Photo 10.5: Riparian flora along one of the creeks I kayaked

The next photo shows an area where a bipedal anomaly visited the creek for a drink, a bath, or perhaps, both. The circle denotes the area of back and forth movement at the river's edge. The line with an arrow shows where the bipedal anomaly walked from and back into the wood line.

Photo 10.6: Area where a bipedal anomaly drank or bathed

I discovered many spots similar to photo 10.6 along the creek. To me, this gives a sense of where bipedal anomalies frequent the creek. Of course, spots of this type are also often found along the rivers. So many possible spots to explore in the vast wilderness within the Northern Highland geographical area! How in the world do we choose which ones to explore first, second, third, and so forth? Good question; and one that I'll address in Chapter 17.

September 14-15: Resting intermixed with slow hikes

The kayak trip was a bit taxing for me due to ongoing heart issues this year; more so than I expected. Therefore, I devoted the next several days to rest, short hikes, and quiet evenings around the campfire. Not wasted time by any means, because each evening was punctuated with nightly visitors. This, of course, returns me to the ideas of *habituation* and *familiarization.*

Habituation has several definitions, dependent upon both the context and the content of discourse. A simple definition that speaks to the way I use the notion in this story is, "Habituation means that when something doesn't pose a threat to our safety, we get used to it."[60] "Getting used to it" addresses the idea of familiarity; ergo, I submit that habituation and familiarity go hand-in-glove.

For years, I have striven to make it plain to sasquatches in our research areas that I do not pose a threat. For instance, one way of showing I'm not a threat is something I learned from someone who had studied with First Nation peoples: I hold my hands out, palms upward to show there is nothing hidden that could cause harm to them. Sounds simple; yet, oftentimes, simple works well. Of course, I have also read that doing this is a way to show submissiveness. I tend to believe that extending my hands outward with the palms up signifies a lack of intent to harm because, to me, submissiveness would require a more elaborate pose. Either way, I can neither empirically prove nor disprove this technique; at least, not at this time.

I also show respect when in the forest and examining footprints. For example, when tracking, I never track farther than necessary to get a sense for the *general direction* they are taking. Once I know the general direction taken, it is a simple matter to peruse my many maps and deduce where they are most likely headed. This approach saves me time and effort, and shows them that while I'm curious about them, I am not trying to *hunt them down*.

I do not engage in calls to them. I do not bait them. Instead, I periodically leave gifts of fruit for them, and, a large percentage of the time, when I leave gifts for them, I do not go back to see if the gifts were taken. I simply trust that if sasquatches do not eat the gifts, other wildlife will do so. I shared one notable exception in Chapter 7 when I related the story about leaving bunches of grapes in a meadow and ultimately received gifts of pin cherry pits in return.

You may also remember me addressing the idea of wood knocking. I do occasionally whack a tree with a large stick or clap a couple of sizeable stones together, and then listen for a response. I have not heard a response to my questionable knocking experiments. I say I engage in this activity *occasionally* because I find it difficult to *not* do this from time-to-time. You know, the inner boy thing sneaking out every now and then. Darn it! No wonder so many folks suggest men never grow up!

Okay, back to the ideas of *habituation* and *familiarization*. The definition I put forth suggests sasquatch's natural curiosity might peak when they do not feel threatened or pursued. While unsubstantiated, I feel that some of the same bipedal anomalies I've met in specific locations in this overall area over the years might recognize me. Perhaps they smell me, distinguish my gait in the woods, or the way I constantly watch all around me. Maybe it's my voice; for I do indeed speak to, at, or with them while I'm wandering through the forest. I speak in my regular voice and make statements or ask questions just as I speak with Julie. Even though I'm quite sure sasquatches do not understand what I'm saying, I have high hopes that Julie does—at least occasionally. I'd ask her but worry about her answer! Or maybe it's my clothing; I wear the same or similar colored clothing while in the wilderness conducting research.

Regardless, I use the terms habituation and familiarization to imply that I have become comfortable with bipedal anomalies and they, in turn, have become comfortable with me. Or, at the very least, they do not feel threatened by me, nor I by them.

It is crucial to note here, however, that no one really knows what sasquatches are, or their origin. Many offer ideas on this; however, I strongly believe that, currently, no one truly knows. This suggests to me that bipedal anomalies, regardless of my level of comfort when near them, might indeed have genetic strains of the "wild beast" within them. For that reason, I remain cautious, and hopeful that I do nothing to inadvertently piss them off. Having something that stands seven to 10 or more feet tall and weighing upwards of 700-1,200 or more pounds mad at me is not something I care to entertain. Nope! Shiver ... shudder!

To wrap up this day, August 15, I wandered outward around camp, looking for evidence of the visitors I hear at night. Sure enough; I found and snapped photos of many footprints. Yet again, the prints were not castable nor were steps and strides measurable. Even so, the many fresh footprints support my belief that sasquatches visit the campsite nearly every—if not every—night. A decent feeling for sure, along with strong support for the ideas of habituation and familiarization.

September 16-25: Slow hikes and kayaking

My body was not as cooperative as my spirit this year. Consequently, this week was one of slow hikes, with a couple of short kayak trips tossed in for variation of flavor. Each hike and kayak trip produced unmistakable evidence of bipedal anomaly activity over a large area. I was also—finally—becoming comfortable with what I needed to do to implement my need to research alone or only with Julie. The time was rapidly approaching when I would act firmly and finally on this need.

I continued to snap photos and wonder at the overall exquisiteness of nature, especially as the leaves were beginning their fall change patterns, turning the forest into a mixture of vibrant reds and golds melded with tantalizing smiles of evergreens or, the green of tamaracks turning swiftly into golden hues of delightful eye candy. Surely, the forest has such incredible tales to tell, and it tells them to those who learn to listen to the breathtaking beauty so often perpetually ignored. I know that I've certainly and deeply enjoyed and appreciated this time alone with the forest and the many stories it shared with me.

September 26-28: Sightseeing and bonding

Maggie rejoined me at camp; however, due to my health issues, she and I embarked upon a bit of sightseeing and bonding. We packed picnic lunches and enjoyed a different gorgeous spot each day. One day we stopped at a small U.S. Fish and Wildlife Service dam.

Following a picnic lunch, Maggie decided to play hide-and-seek, as seen in the following photo. Lucky for me, I found her easily enough peeking mischievously through the trunks of two trees! Guess she wasn't trying overly hard to hide, eh?

Photo 10.7: Maggie playing hide-and-seek near the dam

Photo 10.8 is of a small lake we stopped at in the depths of the wilderness for a few moments just at sunset and before heading back to camp following another quiet day of wandering. I've always treasured capturing clouds in the sky as they reflect in a body of water. This photo was an unexpected treat and pleased us greatly.

Photo 10.8: Sun starting to set over a small lake

I'm not sure what the following is. Even so, we saw this at the edge of the forest a short walk away from the lake seen in photo 10.8. Its beauty and vivid color immediately struck me, so I had to snap a photo.

Photo 10.9: Unknown beauty poking into fall

The next photo was taken as we were breaking camp, readying for the trip home. It captured the fullness of color and crackling stanzas of forest poetry bidding us adieu just before taking down the screened-in gazebo. Remarkable. We are truly blessed each time we visit the wilderness!

Photo 10.10: Smashingly gorgeous color at camp

Chapter 10

[55] DNR, WI. "Wisconsin Department of Natural Resources." Pine and Popple Wild Rivers - Wisconsin DNR. July 29, 2016. Accessed March 17, 2018. http://dnr.wi.gov/topic/Lands/WildRivers/PinePopple/.

[56] Ibid.

[57] Ibid.

[58] Maffei, Massimo E. "Magnetic Field Effects on Plant Growth, Development, and Evolution." Frontiers in Plant Science. August 18, 2014. Accessed March 17, 2018. doi:10.3389/fpls.2014.00445.

[59] Ibid.

[60] Cyprus, Sheri, and Heather Bailey. "What Is Habituation?" WiseGEEK. March 04, 2018. Accessed March 18, 2018. http://www.wisegeekhealth.com/what-is-habituation.htm.

Chapter 11

Habituation and Familiarization Proven — 2015-2016

"A fool thinks himself to be wise, but a wise man knows himself to be a fool."

— William Shakespeare

Two research excursions were made into the wilderness during 2015—one in mid-August, and another in mid-September. Each trip produced its own experiential bipedal anomaly evidence. Habituation and familiarization rapidly moved to the front of my awareness this trip. For instance, bipedal anomalies visited our campsite each night, starting with late evening and lasting long into the night hours.

Part of the time I was alone; part of the time Maggie was with me; and part of the time Julie was with me. Unfortunately, I continued to have heart issues this year, which placed unwelcome restrictions on my physical activity. Bummer! Even so, I accomplished successful research.

The times I was alone are the times when bipedal anomalies interacted most openly with me. This knowledge served as an indicator to me that researching alone has enough value to outweigh potential hazards—if I remained fully aware and always cautious. Further, this idea became both intuitively and openly obvious during my 2017 trip, one that lasted nearly a total of six weeks. Also, it was during the 2017 expedition that habituation and familiarization became the *norm* rather than the unusual. Those amazing stories, though, appear in Chapters 12 and 13.

August 2015 — Maggie Time Before Julie's Arrival

While Maggie was with me, bipedal anomalies visited camp each night. By now, though, they were openly vocalizing while near the campsite. We heard a mixture of grunts, chortles, variously pitched whistles in different tones and lengths, and a sort of gibberish. This sounded as if the sasquatches were communicating amongst themselves.

These vocalizations were interlaced with periodic wood knocking, which, as you now know, I believe is another form of communication for sasquatches. From time-to-time we heard an occasional breaking of small branches. Perhaps an impatient youngster trying to get an adult's attention while they were inspecting the strange humans? Who knows for sure?

On several occasions, sasquatches would gently rock the trailer for a few moments during the night, although never longer than two to five minutes. I'm not sure why they periodically rocked the trailer; however, I believe it an interesting anomaly as well as an indicator of their physical strength. Goodness! They could flip the darn thing over if they chose to do so.

Similar sasquatch visits occurred while Julie was with me. We heard the same types of vocalizations, whistles, gibberish, wood knocking, and light branches breaking. By now, though, this behavior was quite familiar to us—almost to the point that such sounds and interactions with bipedal anomalies had become common. Inspiring movement from the unusual to typical, eh? Wait until Chapters 12 and 13! Seriously! My 2017 experiences run the gamut from implausible to unbelievable, to inexplicable.

Perhaps the sasquatches and me were experiencing nothing more than acceptance of habituation and familiarization. An unexpected embrace, if you will, of something that we once considered quite outside the norm had become a typical and pleasant experience. Regardless, this certainly introduced a different factor into the realm of anomalous experience and cognitive dissonance, although I shan't discuss these thoughts in-depth until later in the book. I will, however, drop a bit of a clue here. I have unmitigated proof that intentional and directed use of coherence brings about amazing sasquatch interactions.

For sure, sasquatches were accepting us differently than during earlier excursions. And we now accepted and welcomed their presence differently than we had in times gone by. Our now familiarity with them was somehow synchronizing with their now familiarity with us. Without intending to throw out a spoiler, this idea expands almost exponentially during my nearly six-week outing in 2017, thereby *obliterating* all walls between the so-called norm and an anomalous experience. I elaborate on these ideas in Chapters 12 and 13, and we are quickly headed there. Intriguing, to say the least!

September 2015 — Maggie and Roger

This trip did not produce research results. Rather, Maggie and I spent time together, wandered around the wilderness in a lackadaisical manner, went thrift shopping, sight-seeing, and visited with friends. We stopped at a couple or so touristy spots for a quick look-see; this fed my recognition that I enjoy the amazing beauty of the great outdoors far more than glitzy tourist-trap attractions.

Oh, we also took a flight over more than 150,000 acres of the country in which we conduct our primary research. Fascinating, revealing, and certainly inspirational! Man, oh man! Maybe one day we can take a helicopter flight over the area and grab several thousand photos. What an incredible treat that would be!

Maggie rarely goes with Julie and I while we wander local rivers and creeks; ergo, she was thrilled to see from the air some of the places Julie and I kayak to and camp overnight. While flying over areas of research, she noted some of the challenges Julie and I face as we move ever more deeply into the vast wilderness of the Chequamegon-Nicolet National Forest.

I had hoped she might take a bit of pity on us once she saw some of these challenges. Instead, I suspect she thought Julie and I were a tad or so nuts. Oh, well. Can't win 'em all, eh? Even so, she was deeply appreciative of the opportunity to fly over the areas we hike and kayak in. It certainly offered her insight that she would otherwise not have had. In addition, the flight gave her a chance to enjoy the overall beauty of a small section of the incredible Northern Highland geographic area. One which she will always remember as time goes on and the research continues.

Photos 11.1 and 11.2 are aerial views of some of the Chequamegon-Nicolet National Forest in which Julie and I have had amazing experiences. Serious yet gorgeous wilderness, I say! Photo 11.1 shows some of the bountiful water supplies available to sasquatches as well as wandering rivers and creeks. It's quite obvious that we have scads of forest in which to wander while conducting our research. There are literally hundreds of thousands of acres we haven't even seen yet, much less had the time to hike!

Photo 11.1: Aerial view — Local river and lakes 1

Photo 11.2: Aerial view — Wilderness and more wilderness

September 2015 — Julie and Roger

Even though Julie and I have a solid two decade plus friendship, we discovered that, under certain circumstances, we might struggle with one another. How unusual to discover that we're human! Regrettably, we had a couple of days that tested the depth and breadth of our friendship. Interestingly to me, this came as a surprise.

I must have been either naïve, or, perhaps, a bit unrealistic to **not** think such a challenge might present itself at some point within the context of our friendship. Regardless, this section points out that neither good intentions nor high expectations are *always* capable of overcoming physical, emotional, or mental exhaustion.

The beauty of the ending of this experience, though, solidifies the idea that a friendship built upon mutual trust, love, and respect can indeed absorb and move through a vast variety of challenges. Here's a shout out to such relationships worldwide!

Three Hard Lessons

Hard Lesson One:

Be sure everything is sorted, packed and ready to move out **early** the next morning when planning an overnight outing on a river or creek. How naturally simple is that?

This day, Julie and I embarked on an overnight kayak jaunt that gave us added data as well as several unwelcome lessons for me. We paddled for a while on one of the local rivers and then along a creek branching off it. We started late because I had not yet completed my towed kayak prep, so we did not reach an area where I had hoped to spend the night. We backtracked, and once we found an alternate location to vacate the creek, we beached our kayaks and moved all equipment to a safe overnight spot.

Unfortunately, we had to complete our overnight camp set-up after dark due to our late departure from the base camp. Thankfully, we carry strap-on headlamps, so the task was easier done than otherwise. We shared camp chores and were soon ready for night. We placed our pup tents and sleeping bags on a thickly moss-covered area, prepared a fire for cooking, ate, and settled in for what we hoped would be an active and telling night.

We were exhausted because of the grueling kayak trip, which included time-consuming paddling, portaging, etc. Our goal was to stay awake in the event a sasquatch or sasquatches chose to interact with us.

I don't recall what time it was, although I suspect it was close to 2:00 a.m. when I heard movement coming from the forest behind us. I quietly asked Julie if she had heard it; she said yes, and it sounded almost as if it were a *duh*! The movement noise ceased of an immediate sudden, with silence pervading the night. The movement noise started up again and, shortly thereafter, we heard a wood knocking sound. An interesting aside here; while I listened for a wood knocking *pattern*, Julie *counted* individual knocks. This, without pre-discussion; again, a tad synergistic, or so it seems to me. And this occurred despite our exhaustion. Cool!

Anyway, I heard a pattern that sounded like knock...knock, knock...knock, knock...knock...knock, and after a moment, it repeated. We were camped no more than 60 yards from the creek and the knocking sound came from behind camp. I estimated the distance to the wood knocking sound as approximately 30 yards. After the second round of knocking on our side, a response wood knocking began from the forest *across* the creek. Hello! So, we were listening to at least two parties that were about 150 yards apart communicating via wood knocking! Another first for sure! Of course, we had no clue how the heck to interpret the communication. We could only guess that the communication was about us.

The wood knocking between the parties on opposite sides of the creek continued intermittently for a couple of minutes. The pattern altered a bit in about the middle of the wood knocking session and then returned to the original one. Julie counted a total of 47 knocks and I noted three distinct patterns. There may have been four different patterns; however, I'm confident of only three.

Although we certainly do not know what the bipedal anomalies were discussing, if such a term is applicable here, it wouldn't surprise me if they were complaining about the damn dumb humans invading their nighttime space along a wilderness creek they claim as their own. At any rate, it was surely an interesting experience, and one we noted for later reporting.

Once the wood knocking ceased, we heard branches breaking behind us. I'm guessing that whatever was there was not overly pleased with results from the wood knocking conversation. The breaking branches seemed the result of a slightly annoyed bipedal anomaly. Oops! Sorry.

Hard Lesson Two:

"Take the time to think things through" *before* acting on what I believe might be a brilliant idea. Our water conveyances at that time were solo sit-in kayaks. That type of kayak does not have a lot of room to hold gear and I haul a *bunch* of equipment into the wilderness. Hence, I had the seeming brilliant idea of making a tow-bar and towing a second kayak behind me to carry equipment and camp gear. Not only was the idea **not** brilliant, it ended up creating many issues readily avoided if proper thought had ruled the day! Photo 11.3 shows me towing the second kayak. Looks quite easy, eh? **Not**!

Photo 11.3: Me towing second kayak — Credit: Julie

Anyway, the morning after the wood knocking session, we decided to leave the damned towed kayak and most of our equipment at the overnight campsite while we moved farther west along the creek to search for, well, whatever we might find. Not having had a lot of rest the night before, and after a challenging day to reach a campsite, we both started this day tired and somewhat disjointed emotionally. Uh oh!

The morning was cloudy and our mood a bit gloomy at the start. Even so, movement along the creek was quiet and ordinary. Hard to stay gloomy, though, while paddling through nature's sublime wonderland. Surely a pleasant respite from towing the second kayak loaded down with stuff and more stuff! Dragging so much equipment into the wilderness is a serious pain—in numerous body parts!

We continued to be amazed at the wild beauty of the area and how it presented sasquatches with a natural refuge in which to move, hide when necessary, eat, drink, bathe when they wished, and quite possibly curse any humans who invaded their sacred home space. We found many areas where sasquatches frequented the creek.

Countless areas were obvious, and we saw a few clear trails leading from the forest to the creek and back. A couple of these trails had seen frequent use, whereas others were used by a single sasquatch. For instance, photo 11.4 shows one spot Julie pointed out to me. A single sasquatch had tramped around there, fussing about in the area for some unknown reason. It may have taken a bath or caught fish, or simply enjoyed splashing around in the river. We will never know why it moved around like it did; only that whatever it was doing, it certainly made a clear impression along the river bank.

Photo 11.4: Footprints at the edge of a creek — Credit: Julie

We also saw several areas along the creek's edge showing a back and forth movement as if the sasquatches who came there shuffled back and forth while doing whatever they needed to do while at the creek bank, then they returned to the forest. We even saw one area where the creek bank was almost level for about 30 or so feet before it climbed sharply upward toward the forest. We named this spot 'Sasquatch Highway' in honor of the many footprints coming and going.

Photo 11.5 shows another area along the creek that grabbed our attention. We stopped to check it out. This one had a trail from and to the forest; the area was relatively well-traveled, yet we saw no distinct or striking prints. Certainly, no castable footprints presented themselves to us. Yet, it was obvious that this was a well-used trail.

Photo 11.5: Path to and from the creek — Credit: Julie

We were surprised at how often sasquatches approached the creek all along the full length of this waterway. But then we weren't, because the area is indeed wild, rarely frequented by humans, and a seeming perfect spot for sasquatches to live without a lot of bother from two-legged creatures like us.

I know! If any were around that morning, we were certainly bothering them, for which we apologize. Well … maybe not. We chose to be there to document sasquatch behavior through observation of general movement or travel patterns along the creek, whether they wanted us there or not. Ethics can cause such brain twisting discomfort at times! Ugh! I wonder if *Excedrin* has a pill for *ethics* headaches?

I cannot emphasize enough, though, the nearly overwhelming beauty of the Chequamegon-Nicolet National Forest, including its many wilderness waterways. Photo 11.6 is one such view. We were looking for a spot to land and briefly check out the forest for possible signs. Sadly, we didn't have time to pursue it. Maybe another year?

Photo 11.6: Looking for a spot to land to check out the forest

We took careful note of areas that we wanted to return to for further exploration, and we surely enjoyed our time along the creek. Now, though, we needed to head back, pick up the towed kayak and equipment (YUK!) and paddle to the boat landing and the truck.

This, we did. We did not, however, yet know what that task would cost us in terms of time, effort, and emotional anguish.

We arrived at the overnight campsite, loaded our equipment and gear on the three kayaks and made a final scout of the campsite before we moved out. Photo 11.7 shows the three kayaks loaded and ready for the trip back to the boat launch. At the time of the photo, it was only partly cloudy. After our final scout of the campsite, though, the clouds were rapidly thickening.

Wouldn't you know it? A brisk breeze came up, which made it even more challenging for us as we would be paddling *into* the wind. Okay. So, it would take us a little longer than we wished. No worries, eh? Except for the fact that both Julie and I were physically, emotionally, and mentally exhausted. And, we certainly did not yet know the extent of our exhaustion or how it would affect our return trip. We found out soon enough, though, and neither of us enjoyed the result.

Photo 11.7: Ready for trip to boat launch — Credit: Julie

Oops. Clouds continued to thicken as we moved along the creek on our trip back to the boat landing. It began to spit a cold drizzle on us, adding to the discomfort of the cool wind. We had to cross four or five beaver dams, which proved to be more of a challenge going back than on our way in. Rats! So now it will take even longer. Ugh! This leads us to the final hard lesson of this section.

Hard Lesson Three:

Don't tow a fully loaded kayak into the wind while it's raining and when you must cross several beaver dams—*especially* if your physical heart is acting up, your blood pressure and heart rate are dangerously low, and you are drained nearly to the point of total physical exhaustion, not to mention emotionally distraught. Geez! A challenge to a friendship? Yup!

Photo 11.8 shows one of the areas we needed to portage. This was a serious physical challenge with a towed kayak! Long … long … trip! Eventually, though, that trip from hell ended and we arrived back at the base camp alive and *almost* fit to be called humans.

Photo 11.8: Area we needed to portage — Credit: Julie

See what I mean about hard lessons? I clearly hadn't honored my heart condition for that trip. I was too stubborn to acknowledge my physical limitations and assuredly paid a price for it—as did Julie. The trip was beyond trying! It taxed us physically, mentally, and emotionally, to the point, at times, where Julie and I ceased communication on our return trip to avoid saying things that could not later be unsaid. Oh, my!

After 20+ years of friendship, this was a literal first for us, although sometimes it is far wiser to hold off on communication until one reaches a space where civility and love reigns supreme, or at least to a degree that aids in toning down potentially ugly discourse. Fortunately, Julie and I remain friends following that unmitigated debacle, so we must have a rather solid foundation to our friendship. I still wonder, though, what the heck those two sasquatches were chatting about during their wood knocking communication that night on the river! How in the world does one learn *sasquatch talk*?

2015 Summary

Following the conclusion of that trip, Julie's time with me was up due to work demands. She headed home, and I spent a couple of days recuperating—for that trip from hell had taken an enormous physical toll on me. Once I was able to get back into the wilderness, I took short hikes only and even managed a single daylong kayak trip. Each hike and the kayak trip produced fruitful information, which was duly recorded for later assessment. Additionally, I reveled in the role of sole researcher. Too, although I had encountered a few serious challenges, none created a crisis while I was alone. This was definitely support for my insistence that I needed to research alone, because if the research went well while I was in a rather poor physical state, it should go much better once I was physically fit. Or so my thinking went; I'm really not interested in running this idea by Maggie or my friends!

The lone kayak trip I embarked upon returned me to the place we had named "Sasquatch Highway." I was alone and carried only necessities, so the trip out was not too taxing; rather, it was quite pleasant. I beached the kayak, climbed a steep hill, and was amazed at the bipedal anomaly signs I saw. Wow! Wish Julie had seen this; she'd have been elated!

Anyway, I spent the day hiking along the creek and back into the wilderness. Sasquatch signs abounded! The scenery was spectacular, the forest symphony of sound wafted over me, and I felt a sense of quiescent, soothing, and healing electromagnetic waveforms gently flow around and through me. Imagination gone wild? Doesn't matter!

Whatever it was, it surely helped me feel better! In fact, I felt so relaxed I napped for an hour amid this immensely gorgeously early fall wilderness. What a delightful and unexpected treat!

Son-in-law Teaches me to Kayak

Another brief aside: I've mentioned throughout the book how none of this extraordinary research would be possible without aid from many people. Along that vein, it's way past time for me to acknowledge a gracious son-in-law. He is an avid kayaker and he and his friends paddle and camp along rivers throughout the state of Michigan each summer.

This young man taught me to kayak; he is an excellent teacher, thoughtful, patient, and respectful. He and a daughter loaned me kayaks for many of our trips into the wilderness. Without them, Julie and I would have been land-bound, which, of course, would have limited our research, substantively altering end results.

We would have missed countless of the wonderful experiences we had when we wandered local rivers and creeks. So many, in fact, our research and this story of sasquatch and me would be far different than presented here. Anyway, here's a grateful shout-out to a cool and sometimes "fancy" son-in-law. Thanks bunches, Son. You are greatly appreciated and loved. In fact, you rock!

The 2015 research trips ended with me feeling that they were successful. All arrived home alive, intact, and ready to begin evaluating the gathered data, which prompted planning for the next round in 2016. Due to my time alone this year, I knew that future trips would be mostly me alone or only with Julie. The decision was made—finally—and I felt as if an albatross had been lifted from my shoulders.

Little did any of us know that fate would cast an unexpected, dreadful obstacle directly into our path in the fall of 2016. We never know what might be around the next corner! Lesson? Enjoy today with exuberance!

2016 Was a Bust for Research

Maggie had brain aneurysm surgery in September 2016. For sure and certain, our world fell apart for a stretch of time. She continues her recovery to this day, and we hold high hopes for her longterm well-being. For example, she was able to take a couple of short hikes in 2017, for which we are deeply grateful. More about that later. Due to her medical condition, no research was conducted in 2016.

Maggie's neuro surgeon did, however, grant permission for her to travel to our research area in late October 2016 as long as her physical activities were minimal, and someone was **always** with her. We rented a cabin on a lake for a week from Mark and Julia, who also own the Little Pine Motel in Hiles, Wisc., and enjoyed time with friends, including a lovely, albeit chilly, late afternoon pontoon ride around the lake with Brushtracker and Dragonfly, who had moved from Michigan to Wisconsin.

I decided to take a day and check out the research area. Mark, Brushtracker, and Dragonfly visited with Maggie while I was gone. I drove around much of the day checking various sites. Per usual, I noted a lot of activity.

Late in the afternoon I chose to hike an area where I had not been before. It was along a creek with forest and thick brush closely encroaching upon, and valiantly clinging to, the banks of the creek. I was excited to check it out.

The forest here was a mixture of various pines and hardwoods and was indeed stunning in its overall beauty. Numerous sasquatch signs prevailed, and I eagerly followed them, wondering where they might lead and if they would end at the creek. After walking about 150 yards into the forest, of an immediate sudden, 10 Black-capped chickadees began swirling about my head—chirping loudly and flapping their wings so rapidly that it was almost like they wanted to *wing* me. Say what? What'd I do to them?

When the 10 chickadees began circling my head and chirping loudly, my first thought was, "This is seriously cool!" As I continued to slowly walk forward, however, the chickadees began to chirp even more loudly while frantically flapping their wings as they circled me, apparently trying to grab my attention.

To suggest that the experience I was having with the ten chickadees on this hike was highly unusual is a clear understatement. I had no idea why they were *wing flapping* around me in such a harried manner!

Black-capped Chickadee Backstory

I need to take a moment to give you a chickadee backstory here. All summer long at home, I had an interesting situation with two chickadees. When I would step outside the garage door into our back yard, either one or two chickadees would fly toward me and circle me while I was walking around the yard. If I stopped just outside the door, one of the chickadees would actually light on my left shoulder, chirp quietly into my ear for a few seconds, then fly away. Clearly, that was a delightful experience that I greatly enjoyed!

I would stop, and their circling became slower, their chirping quieter. When I started to walk again, they would once more flap their wings feverishly and chirp loudly. I finally decided that they were trying to inform me of something. I stopped, grabbed my binoculars, and began a 360-degree eyeball search all around me. I may be a tad slow at times; however, I'm not a **no** learner!

Oh my gosh! About 60 or 65 yards directly ahead of me was a bear den along the creek. It was approximately 50 feet up the bank from the creek. At the opening of the den and looking directly at me was a

standing black bear! And that sucker eyeballed me like it was either hungry or surly! I estimated its weight to be around 300-325 pounds.

I looked at the 10 chickadees, who were quietly circling my head at that time, offered them a grateful thank you as well as a loving blessing, and headed back to my vehicle. Interesting experience, eh? I might be wrong in my assessment here; however, it seems to me that the chickadees were warning me **not** to continue in the direction I was headed. Duh! The whole experience was indeed both thought-provoking and stimulating. I'll not forget it any time soon.

Chapter 12

Mind Blowing Encounters Part 1 — 2017

"Patience, persistence, and perspiration make an unbeatable combination for success."

— Napoleon Hill

Our research efforts in 2017 literally shattered all earlier expectations, some of which were quite high at the start—or so I thought. I had phenomenal experiences; some stretching the realm of believability nearly to a breaking point. Even so, I propose that the boundaries surrounding incredulity can be broken much as the sound barrier was broken unexpectedly on Oct. 14, 1947, by U.S. Air Force Captain Chuck Yeager. Of course, there are some who claim Yeager was not the first to break the sound barrier; rather, that the sound barrier had been broken by others prior to his flight. This suggests that controversy surrounds that momentous moment. Sadly, it does.

Controversy also persistently assails the realm within which sasquatch phenomena exists. In fact, controversy seems to be a mainstay within the sasquatch community because, regardless of the type of evidence produced to support sasquatch claims, I am not aware of a single instance when the evidence was not refuted, claimed to be fraudulent, or otherwise maligned. I mention this here only to remind you that whatever I propose as supportive evidence of my interactions with sasquatches, I am fully aware that many will refute and/or malign it. No worries! I know what I've seen. I know what I've experienced, and I certainly believe what I've both heard and seen. Frankly, I know and am comfortable with my sasquatch reality and am not overly concerned what others' think.

I stated at the start of this book that my intention in sharing this story is not in any way related to attempts at proving sasquatch is real. My experiences with sasquatches over the years has solidified my belief that they are real. I also have no intention to try and convince others that what I believe is real. Rather, I merely wish to share my story of nearly two decades of sasquatch research so that those who hold a similar belief might add yet another possible tool to assess their experiences.

That tool is: Understanding Earth *magnetism* as it interacts with *geomagnetism* while *coherence* produces startling sasquatch encounters. Seriously! Keep reading.

2017 showed repeatedly that habituation and familiarization as used in this book is, indeed, valid. The total time I spent in the Northern Highland geographical area in 2017 was just shy of six weeks. Frankly, that length of time in the wilderness in a minimal base camp can a bit grueling, to say the least, and a tad overwhelming at times, to say more. Despite periodic discomforts, though, results from the trip were amazingly fruitful, astonishing, quite telling, and very satisfying.

To say our experiences this year were remarkable is clearly more than an understatement. For instance, I saw **three** full-bodied, walking sasquatches in **three** different locations; one while camping overnight along a creek, one while hiking, and another one at my base camp. I share sightings details throughout Chapters 12-13. Each of the three I saw were undoubtedly female. Moreover, other unusual experiences occurred near the end of my research time in 2017. I discuss those unique, mind-blowing encounters in Chapter 13.

For now, though, let me remind you that I define *habituation* and *familiarity* in this book as, "when something doesn't pose a threat to our safety, we get used to it."[61] I believe the 2017 research trip exemplified this concept for me; that is, I became quite accustomed to interacting with sasquatches dilly-dallying around my base camp and did not feel threatened by them. Neither was I startled when I saw one while I was hiking or one during an overnight camping trip in the wilderness. In turn, it seemed as if the sasquatches who repeatedly visited my base camp became quite familiar with me, to the point that they did not feel threatened by me or what I was doing.

Once again, reason dictates that considering their size and physical prowess, I would likely feel threatened by them much more quickly than they would me. Even so, not feeling threatened melded with familiarity seems to represent the way we felt about each other. In short, I suspect a mutual sense of familiarity contributed to the amazing interactions I had with them throughout the 2017 research trip.

A word of caution here. Familiarity should not, under any circumstances, detract from the reality that sasquatches live in the wild and they most likely have a "wild streak" embedded within their nature. One which contributes greatly to their ability to survive in a raw and oftentimes unfriendly wilderness. As such, it is of utmost importance to always remember the potential danger one could put oneself in through intentional, close interaction with one or more sasquatches. I honor this warning dutifully, particularly when I'm within 25-35 feet of them, which I certainly was near the end of my 2017 research trip! Really!

September 14-19: Setup camp and early surprises

I spent September 14-16 setting up camp and preparing for an overnight canoe trip. I heard visitors moving around just outside camp on the evenings of September 15-16. The morning of September 16, however, presented me with an unexpected surprise.

Upon arising and stepping outside the trailer, I noticed footprints between the back of the truck and the trailer. Further looking showed the truck tailgate had fingerprints in the dust. These were clearly visible as seen in photos 12.1 and 12.2. It appears to me as if a hand had lightly rested against the dusty tailgate; twice! One on the left side of the tailgate, and one on the right side.

Okay. Apparently, sasquatches are curious enough about me to move closely into my camp area. I always park the truck in front of the trailer tongue, with only a few feet between it and the trailer. Therefore, the truck tailgate is about 4-5 feet in front of the trailer tongue. This means that when a sasquatch leans against the front of the trailer, or walks between the truck and the trailer, it is literally within two to four feet of me while I'm sleeping. Reminder: The trailer front wall is barely 1½ inches thick, with much of that air. Think about it!

Photo 12.1: Fingerprints on truck tailgate (left side of truck)

Photo 12.2: Fingerprints on truck tailgate (right side of truck)

Footprints appeared near the truck tailgate. They measured 25 inches long, 6½ inches at the heel, and 5¾ inches at the midfoot. Neither step nor stride were measurable.

I spent September 17-18 driving for hours around the entire research area, stopping from time to time to check for sasquatch activity. I was both pleased and somewhat startled at the amount of activity I found and decided then and there this was likely to be an exciting trip. I wasn't wrong!

I planned to leave for a four-day, three-night canoe trip on September 19. Inclement weather ensued, so I spent time near camp for an extra day. I had nocturnal visitors wandering around the outer edge of my campsite each night. Footprints found each morning confirmed that sasquatches were showing an increasing interest in my activities. I do not know why; I know only that several sasquatches visited my campsite each night.

I would sit outside at night with a small campfire burning brightly and speak with them as if I were speaking to a friend. I spoke about the weather, the fall season with its changing colors, and other things. They did not respond to my voice other than to move about to let me know they were around. By the way, I avoided discussions on politics knowing how tenuous emotions are about the divisive nature of today's political drama. I certainly didn't want to piss them off if I inadvertently said something they didn't like! Hmm! Wonder if they were registered to vote?

Regardless, they moved through the forest and all-around camp each night *without stealth*, which told me they wanted me to know they were checking me out. Again, I confirmed their presence each following morning by observing footprints.

September 20-21: Overnight kayak trip

On September 20, I loaded up overnight camp gear and headed to the river launch. I canoed the local river and a creek. I at once noticed a rather distinct difference from earlier years. First, the local river was high again. Second, the creek I entered off the river had altered course—quite drastically. This was not only a surprise, it was somewhat befuddling. I had no clue then what might have caused the creek to alter course in that manner. I only knew that it had done so since the 2015 research trip.

As I continued to explore the creek, I found that beavers had built a huge dam across it. Although Julie and I had seen and crossed many beaver dams over the years, none were as large as this one. This dam was so large that the creek had backed up! In turn, this altered both the landscape as well as the directional flow of the creek, both before and after the beaver dam. An astonishing find to be sure! I did not cross the large beaver dam at that time; rather, I turned around and paddled back to the spot where Julie and I had camped overnight during our 2015 research trip.

My first night out was uneventful. I set up camp and wandered the woods looking for sasquatch signs. Although I found many areas where fresh signs showed sasquatch activity, nothing struck me at that time as very notable.

When we camp on a river or creek bank, the camp is minimal; you can see this from photo 12.3.

Photo 12.3: Overnight camp site for September 20-21, 2017

For orientation purposes, the back of the pup tent is pointing north. The creek lies approximately 60-65 yards to the south and winds around so it lies about 100 yards to the east (right side of the tent). I had packed enough sandwiches that I didn't need to dig a firepit for cooking—this was truly a minimalist camp. Gotta love PB&J along with crackers and cheese!

The tree to the right of the pup tent held a small lantern, while the canoe and supplies were just south of the tent. I turn my canoe upside down and place gear under it for protection. This time, I tied the canoe between two trees because the wind had increased, and I didn't want it or my gear to move surreptitiously about during the night while the wind danced playfully—or wickedly—through camp. Had I packed food to cook over a fire, I would have hung it in a tree.

The second day in the creek camp found me once again wandering north into the woods. I covered a large half-circle around camp that day.

Recall that the creek lies south and east of the pup tent, so the half-circle I hiked was a bit northeast, north, west, then southeast back to the creek. Julie and I had not hiked this area when we camped here during the 2015 research trip. This hike was both enjoyable and informative. Julie would have enjoyed it!

I hiked at least a half-mile into the woods as I made my half-circle. Interestingly, I found quite a few sasquatch trails moving west and northwest. I say this was interesting because I saw only three suggestions of single sasquatches moving in the opposite direction. Okay. Common sense dictated there had to be a reason that most of the signs showed sasquatch movement west-northwest as opposed to typical wanderings in nearly all directions.

Of course, I'll be the first to admit that I don't *always* act with common sense—or even know what it is—if I'm in the middle of an exciting hike or canoe trip. This time, though, I did employ common sense, which means I continued the hike with a clear eye for safety, orientation, and documentation for later review. That review time is now as I write this, which shows why taking the time to faithfully document is so important.

Without documentation I would be assigned to memory only, which, as I'm sure most of the readers of this material know, is not always what we wish it to be. Nonetheless, all that to say I would have forgotten to include part of the data gleaned from that hike had I **not** documented along the way.

For example, I wanted to comment on *regional migration* and would have forgotten to do so if I did not have my notes. I believe my experience with regional migration is noteworthy enough to place my thoughts on it in a separate section, which comes up shortly.

Very Early Morning September 22: Return to camp

The second night out proved to be a bit more eventful, for this was the night I heard **four** sasquatches and saw **one** adult full-body female sasquatch as it walked along the creek not far from my pup tent. I'm confident she was female because I had a profile view of her and saw her breasts moving in a natural way. They seemed symmetrical with the rest of her body, not ponderous as I've often read. Wait—I'm getting a bit ahead of myself here.

Before seeing the female sasquatch walk across my view, I heard two distinct yet resounding splashes in the creek; they sounded about 20-30 yards apart. My guess is that the one I saw was walking to meet a second one. I suspect the second one was a male for two reasons: first, because its splashing and movement was much louder than the female's, suggesting a larger, stronger body, and second, in my experience, female sasquatches almost always seem to travel with a male or a male and a youth. Of course, I have also seen signs where I presumed a female sasquatch was traveling alone with a youth. I find it hard to imagine, though, that a male sasquatch was **not** nearby those times; I say this because I believe that sasquatches are family oriented.

Anyway, night came, and I watched and listened. The time was 2:45 a.m. The sky was cloudy, a breeze of about eight miles per hour was blowing, and it was a new moon, which made the wilderness quite dark. I had also noticed my barometer, which prompted a bit of concern for me about how rapidly it was dropping. At this point, however, there was no rain. The clouds made the night seem even darker than otherwise, yet the semi-open area between me and the creek had enough light for me to see a sasquatch walk across my view.

I quickly turned my head when I heard walking noises north of me, which was behind the pup tent. After approximately five minutes of keen listening, I heard what sounded like communication between two entities; I assumed they were two additional sasquatches. One voice had a lower pitch than the other, which caused me to believe another male and female were heading south toward the creek—and my camp was between them and the creek.

All at once, the walking ceased, and the lower-pitched voice became silent. The higher pitched voice continued for a moment longer. The lower-pitched voice made a clear, distinctive guttural sound and the higher-pitched voice instantly ceased. I then heard a large branch break, accompanied by another deep-pitched low guttural sound, albeit not a growl.

Although I'm not certain, the guttural sound made me believe that whatever was out there might be a little annoyed at my presence. I then heard these two bipedal anomalies turn west-northwest, trending in a half-circle pattern around my camp, similar to the one I had taken the day before as I hiked the area. They seemed to be circling my camp while heading toward the creek. I estimated their distance from me at that point to be about 20-25 yards, which means they were quite close—or so it seemed to me. My heart was beating fast and my adrenaline was rising. I saw only vague, ethereal shadows of these two; therefore, I did not count them as a clear visual encounter with sasquatches. For your information, I do not consider observation of *ethereal shadows* sufficient evidence to claim such observations as me *seeing* sasquatches. I must be able to describe specific physical features before I claim them as being *seen* by me.

During the few minutes this occurred, I continued to hear splashing in the creek. Again, the two I heard coming from the north seemed to be circling my camp as they headed toward the creek through the thick woods around me. As they did this, I heard several medium-pitched guttural sounds which, I believe, came from a female. Shortly after that, the splashing in the creek to the southwest ceased and I heard brush crackling and footsteps approaching the camp from the west-southwest.

Turning my attention to that direction, I saw a shadowy figure appear from the bushy area into the clearing between the creek and me. As this figure moved forward, it paused for a mere moment in the semi-clear area of light cast by my small lantern, looked directly at my pup tent in which I was sitting, and then walked straight east toward what I believe was a male sasquatch.

Again, while the brush was quite thick along the creek, the wooded area between me and the barrier was thin; hence, I had a clear view of the zone between the barrier along the creek and me. The female bipedal anomaly that had been in the creek walked through this clear area toward the other sasquatch, who continued making splashing sounds. The dank smell of the river was quite strong and seemed to be emanating off her. Guess I don't want to bathe in this creek!

I saw her profile quite plainly, and confirmed she was, in fact, female. She did not act disconcerted in the least, despite being so close to me. In fact, she walked rather nonchalantly across my vision as if I weren't there. At one point, she was about 45-50 feet from me as she crossed in front of my camp. She then moved back toward the creek and I heard a brief communication between her and the one toward whom she walked.

At this point, I heard her move into the creek with the other one. Simultaneously, I heard the two who had walked around my camp also move into the creek, about 35-40 yards west of the others. I then heard loud splashing sounds as the four of them crossed the creek toward the wilderness on the southern side of the creek. The time was now 3:07 a.m. so the entire episode with these four sasquatches lasted about 22 minutes. The female sasquatch was within my purview of vision for about 2½ minutes. Her body hair was about four inches long, and I judged her height between 6 feet 11 inches and 7 feet two inches. Her body was muscular and her movements smooth and graceful. I was unable to discern facial features or other details. This, then, became the **sixth** walking sasquatch I'd seen since 1999.

The part that amazed me most about this encounter is the way she seemed comfortable being so close to me. Not only that, none of the four displayed angst in any way. Although the one behind my camp may have been a bit annoyed at my presence at the outset, the sasquatch with it seemed to accept me being there, and they went on about their nightly business without disturbing me or grumbling to any further extent.

I took time to document this experience, and then slept, awaking at 6:30 a.m. to a rather brisk wind and heavy, low clouds. I carry a walkie-talkie with a weather radio while I'm in the wilderness. Looking at the sky caused me to listen to the weather forecast, and I was surprised to immediately hear weather warnings. According to the alert, a severe storm was rapidly moving into the area. The warning was for high winds and hail, and it informed all who were within the path of the storm to seek immediate shelter. Scat!

At best I was 1½ hours from the boat launch and the truck. I broke camp at once and quickly loaded everything into my canoe. As I entered the creek, I realized I would be paddling into the wind, which would slow me down. This, of course, brought back memories from the 2015 trip from hell that Julie and I had "enjoyed" together. Not!

I had not been on the creek or river during a strong storm before and had no idea how tumultuous it could become. It began to rain lightly about 10 minutes after I left the campsite. I had my raingear on, so was as prepared as possible for the drizzle. Before long the drizzle turned into a rousing downpour driven by high winds.

The journey back to the boat launch was arduous and, at times, a bit dangerous. Who knew that the darn river could get a tad or so angry and punish the canoe with high waves? I learned this quickly, though, and adhered to strict safety procedures the whole way. I'm convinced that following safety procedures prevented me from experiencing close calls on the way back to the boat launch. And, I assure you, I will do all in my power to avoid such a canoe trip ever again!

I sure was happy, though, when I finally pulled into the boat launch. The journey took me nearly three hours! As I beached the canoe, it began to hail; small hail at that time, yet it stung exposed skin. I rapidly loaded all gear onto the truck, covered it, and drove safely back to camp.

Dang! The weather advisory sure was right about wind and hail. I had to dodge downed trees on my way back to camp and low spots in the road were rapidly flooding over. Anyway, I made it safely back to camp, covered the truck with a tarp to protect it from hail, and huddled in the trailer until the storm blew over. I would not have wanted to be in my pup tent during the hail storm, or on the water with waves crashing into the canoe while large hail lashed at any exposed skin! Thank goodness for the weather radio … a clear and mandatory safety device whilst trekking deeply through the wilderness.

Regardless, the two nights out were exciting, informative, and provided me with added research data, not to mention a personal and up-close visual encounter with a sasquatch. Indubitably well worth the trip! Next, I speak a bit about what I believe might be a form of sasquatch regional migration. This, because of the way the creek had altered direction and, was backed up by a large beaver dam about an hour west of the overnight camp.

Regional Migration

To start, recall that the creek I was on for my overnight camping trip had altered course due to construction of a large beaver dam across the width of the creek. Also, the flow of the creek following the beaver dam was both directionally different and contained much less water than previous years. In short, the water supply in this area had changed a good bit. Considering the change in the area's water supply, combined with my hiking this zone, I had noticed a clear and unusual movement of sasquatches toward the west-northwest. I decided to pursue this mystery.

Perusal of my maps showed that the creek naturally lessened as it flowed westward. Adding the change in quantity of water between the beaver dam and the spot where the creek naturally lessened suggested that the water supply for sasquatches and other wildlife also lessened within this multi-mile zone. I suspected this meant that the potential migration I'd noted earlier might be occurring as sasquatches sought alternate water supplies. Following this insight, I drove to where the creek nearly ended and … Voila! Sure enough, *myriad* tracks confirmed my suspicions; that is, a regional migration seemed in progress.

My next step was to revisit my maps. I discovered a variety of potential water sources in a north-northwest direction and decided to confirm my suspicions by hiking to a new area of wilderness. I lack the linguistic form to adequately describe the quantity of footprints I found. Stated simply, unbelievable!

The prints were so abundant that it was almost as if I was following a major "Sasquatch Highway." There were many more signs here than those Julie and I had found along a creek in 2015. I'll have to come up with a different name for the creek highway—maybe "Sasquatch Trail." After years of private sasquatch research, I had not seen so many prints in one location—with most of them moving the same direction—directly toward ample water supplies. I was so astounded that I had to sit on a fallen tree to ponder what I was seeing. Literally. My mind had difficulty grasping, and then assimilating, the evidence in front of me. How deliciously droll!

The meadow in which I discovered the footprints was, per usual for the Northern Highland geographical area, open and beautiful. At the time I was there, it was quiet, serene, and calming, other than seeing the myriad prints. The width of Sasquatch Highway was 28 paces, or about 84 feet wide! Even as I write this, I find it challenging to grasp the significance of my words. As a reminder though, I mentioned at the start of this chapter that 2017 was filled with mind-blowing encounters and experiences. This instance certainly qualified as one of them. At any rate, this undoubtedly seemed to fit quite well within the premise of regional migration as I understand it, and this regional migration appeared driven by sasquatches seeking new or different water sources. Now, back to the ongoing story.

September 23-30: Inclement weather

Cold, wet weather settled in for the long haul. The river quickly became too high to pass under the bridge; ergo, water travel was postponed. On days of light rain, I continued hiking various locations, always on the lookout for added evidence about sasquatch behavior. Storm days also found me spending more time at camp or in the nearby woods. Rain or shine, though, sasquatches stopped by camp each night; and each night they moved closer to being within visible range. Habituation and familiarization? Maybe. Nah, undeniably!

October 1-13: Maggie joined me for a few days

Maggie joined me for a few days; we arrived at camp at 8 p.m. of an evening and I unloaded items from the truck. At one point, I leaned my back against the front of the trailer for a short break and simply looked around camp and into the darkness of the woods. I neither saw nor heard anything. Yet, I felt as if something was watching me. Once again, this is not an uncommon feeling in sasquatch country; ergo, no worries.

We had a late dinner that night and went to bed around 10:30 p.m. At around 1:15 a.m. we awoke to the trailer gently rocking; this lasted for about five minutes, which, by the way, was the amount of time I had leaned against the front of the trailer during my earlier rest break. Hmm!

Lo and behold, the following morning we discovered tracks on either side of the trailer tongue. The tracks were facing away from the trailer just as mine had the night before. Apparently, *two* bipedal anomalies mimicked my leaning against the front of the trailer. As they leaned against the trailer, the unit rocked. They must have shifted positions as they stood there because the trailer continued rocking for about five minutes. This means that they stayed in that position for about the same amount of time I had the evening before during my rest break.

I'm convinced they meant no harm because, had they wanted to, they could easily have flipped the trailer over. Rather, they simply mimicked my behavior! This was added evidence to me and was quite exciting. Further, I believe this speaks once again to the ideas of habituation and familiarization. So much so that this incident qualifies for inclusion into another proposal:

Proposal 6:

> Sasquatches have strong observational skills, an aptitude to assess time in some manner, the skill to mimic human behavior, and can use discernment.

While I'm addressing habituation and familiarization, an interesting study published in the *Journal of Veterinary Behavior* in 2013 concluded that it is not uncommon for domesticated cats living with humans to adapt their lifestyle with their owner.[62]

Obviously, bipedal anomalies are not cats. Of course, I have read several accounts suggesting that sasquatches have a bit of an affinity for cats and may sometimes keep them as pets. Regardless, the idea that animals, including horse and sheep, adapt their lifestyles with humans opens a wee window into the possibility that sasquatches may engage in similar behavior if they gain enough familiarity with humans. Or so it seems to me. Little by little, sasquatch evidence continues to seep into our awareness; with enough evidence, we can begin to draw accurate conclusions.

In addition, results of another fascinating study linking the energy of a human heart with the energy of a pet's heart by the HeartMath Institute was published in 2003. Seriously! For instance, HeartMath lead researcher, Rollin McCraty claims that a type of synchronization is possible between humans and pets.[63]

In that study, basically, a 15-year old boy and his pet dog were tested. McCraty placed one Holter cardiac monitor on the boy and another on his dog. The dog was placed in a lab, and when the boy entered the lab and sat down, he was instructed to feel love for the dog. An amazing shift in the heart rhythms of *both* the boy and the dog occurred throughout the time the boy consciously felt feelings of love toward the dog, leading to a conjoined state of coherence.[64] Wow!

Ah, yes. Yet another mention of coherence, although this time, one between a boy and his dog. Finally, I believe heart energy is heart energy whether in humans or all other living organisms; therefore, coherence *between* humans and sasquatches must also be possible. Yes, I truly believe this, and address the idea more in Chapters 13, 15, and 17.

Excitingly, piece by piece, this story is heading toward experiences that continue to astound me even today. I'm not sure how anyone could have these or similar experiences and not get pumped up about our enigmatic sasquatches. Really! Oh well, each to their own, and for us this quest takes first place along our awareness spectrum.

Proposal 7:

> It is possible for a conjoined state of coherence to occur between a human and a sasquatch.

Of course, I also had to take Maggie to see Sasquatch Highway. The hike was only about 300 yards from the parked truck, and she managed it quite well. She was as impressed and thunderstruck as I had been upon my first visit. Who wouldn't be? Not only that, this was the first short hike she had been able to take following her brain aneurism surgery. We also had the treat of a buck and doe jumping up eight feet ahead of us and loping off into the wilderness. We were thrilled!

Now, before delving into experiences that happened during the rest of this trip, I need to introduce the concept of geomagnetic storms and why the topic is important to this discussion. To begin, I have strong evidence that geomagnetic storms played a noteworthy—if not dominant—role throughout my 2017 research. You will see what I mean about that as we move through Chapters 13-16.

Introduction to Geomagnetic Storms

This section introduces why I believe tracking geomagnetic storms are essential to this story. Earth experienced many geomagnetic storms during my September-October 2017 research trip. That is, Earth's magnetosphere was either unsettled or otherwise disturbed for many days, sometimes, seriously disturbed.

Geomagnetic storms affect Earth's ground magnetism, which is susceptible to rapid changes throughout the period of storms, and sometimes a bit before as well as a bit after. Becoming aware of this information is significant to the ongoing story of sasquatch and me. This statement becomes clearer as we delve into Chapter 13; further, I address geomagnetic storms in more depth in Chapter 15.

You may recall from earlier discussions that geomagnetic storms are rated according to the National Aeronautics and Space Administration's (NASAs) so-called *Kp index scale*. The Kp index scale is rated from Kp 1-9, with Kp 9 standing for an extreme geomagnetic storm. The Kp index is basically a measure for "the rate of change of magnetic fields near Earth."[65] Recall that my 90s heart-head research detailed information about how Earth's changing magnetism affects the human body, particularly, the autonomic nervous system. Also, the autonomic nervous system is key to the act of intentional coherence.

When the Kp index hits 5 or above (geomagnetic storms), Earth's magnetosphere is disturbed much more than when the Kp index is 4 (which stands for an unsettled condition) or Kp 3 or below, which indicates there is no storm or disturbance. Geomagnetic storms are also rated from G0-G5, with G5 classified as the most severe.

Table 12.1 associates storm strength, G0-G5, with the Kp index 1-9. This is basic geomagnetic storm information and useful in understanding later discussions. Of course, it is also important to note that when the Kp index is zero for a period of 12 or more hours, a phenomenon known as cosmic rays occurs. More on that later as well.

Table 12.1: Geo Storm Indicators, Kp, and Storm Strength

Geomagnetic Storm Indicator	Kp Index	Geomagnetic Storm Strength
G0	Kp 1-3	No magnetosphere disturbance
G0	Kp 4	**Unsettled** magnetosphere
G1	Kp 5	**Minor** geomagnetic storm
G2	Kp 6	**Moderate** geomagnetic storm
G3	Kp 7	**Strong** geomagnetic storm
G4	Kp 8	**Severe** geomagnetic storm
G5	Kp 9	**Extreme** geomagnetic storm

Photo 12.5 is a NASA Goddard Space Center depiction of Earth's magnetosphere.

Photo 12.4: Magnetosphere — Credit: Goddard Space Center

Earth's magnetosphere, while quite complex in nature and function, may be viewed simply as a type of *shield* or non-uniform *bubble* surrounding Earth. This shield protects Earth from damaging solar wind effects. *Pys.org* discusses Earth's magnetosphere. They wrote:

> Enveloping our planet and protecting us from the fury of the sun is a giant bubble of magnetism called the magnetosphere… Without the magnetosphere, the relentless action of these solar particles could strip the earth of its protective layers, which shield us from the sun's ultraviolet radiation.[66]

Eftyhia Zesta, Chief of Geospace Physics Laboratory, NASA, wrote, "The earth's magnetosphere absorbs the incoming energy from the solar wind, and explosively releases that energy in the form of geomagnetic storms and substorms."[67] Most scientists agree that the largest part of Earth's magnetic field "… is produced by a huge, dynamo-like, rotating solid core of iron-nickel alloy at the center of the earth. That core is spherical, with a radius of 1220 km [~758 miles]. It effectively acts like a huge bar magnet with an axis."[68] The idea of Earth acting as a bar magnet is addressed in Chapter 15.

Table 12.2 shows the research days in September-October 2017 that had geomagnetic storms or an unsettled geomagnetic field.[69] Each page of this story now takes us closer to understanding how I created a proposal associating Earth magnetism, geomagnetism, biomagnetism, solar storms, and the Laurentian Shield of the Northern Highland geographical area with sasquatches.

Recall that I introduced the *Laurentian Shield* in Chapter 2 during a discussion on extremely low frequency (ELF). Recall, too, that the U.S. Navy made use of the natural properties of the Laurentian Shield to build an antenna system capable of communicating with submarines in oceans around the world.

I revisit the properties of the Laurentian Shield in Chapter 15; for now, take a peek at table 12.2 to see how many days of geomagnetic storms or magnetosphere disturbances occurred during my 2017 research trip. You will see that there were *four* days of storms rated G6, *one* day of a storm rated G6-, *four* days of geomagnetic storms rated G5, *four* days of storms rated G5+, *one* day of a storm rated G4+, *three* days of storms rated G4, and *one* day a storm was rated G4-.

This totals a remarkable **13 days** of geomagnetic storms and **five days** where Earth's magnetosphere was unsettled. This is an unusual—if not unprecedented—stretch of persistent geomagnetic disturbances. Also, this data does not reflect the **near** "kill shot" Earth received on September 17, 2017! Oh, my! That's another thrilling story; it, though, it is not included here.

Table 12.2: Kp index in Sept. and Oct. Geomagnetic storms

September 12.18, 2017		
Date	**Kp Index**	**Storm Strength**
September 12	Kp 5+	**G1 Storm**
Date	**Kp Index**	**Storm Strength**
September 13	Kp 5+	G1 Storm
September 14	Kp 5+	G1 Storm
September 15	Kp 6-	G1 Storm
September 16	Kp 5+	G1 Storm
September 17	Kp 4	G0 Unsettled
September 18	Kp 4+	G0 Unsettled
September 27-30, 2017		
September 27	Kp 6	G2 Storm
September 28	Kp 6	G2 Storm
September 29	Kp 6	G2 Storm
September 30	Kp 4	G0 Unsettled
October 02 and 11-19, 2017		
October 02	Kp 4	G0 Unsettled
October 11-15, 2017		
October 11	Kp 5	G1 Storm
October 12	Kp 5	G1 Storm

October 13	Kp 6	G2 Storm
October 14	Kp 5	G1 Storm
October 15	Kp 5	G1 Storm
A 5-day stretch of polar geomagnetic storms ended as Earth exited a fast-moving stream of solar wind.		
October 20, 2017		
Date	**Kp Index**	**Storm Strength**
October 20	Kp 4	G0 Unsettled

In the end, a startling total of **18 days** where Earth's magnetosphere experienced various magnitudes of disturbances occurred during my 2017 research trip. This many geomagnetic storms in such a brief period is, indeed, extraordinary. With Earth's geomagnetic field so active during that period, the prospect for deliberate use of changing magnetic energies for research became real and practical. In fact, evidence supports my premise that geomagnetic storms create opportunities for intentional and directed interaction of magnetic energies between humans and sasquatches. Exciting, eh? This leads to another proposal:

Proposal 8:

> Geomagnetic storms offer sasquatch researchers opportunities to engage in deliberate use of changing Earth and geomagnetic energies while observing and intentionally interacting with sasquatches.

Just wait until you jump into Chapter 13, which is up next. The idea of *extraordinary* takes on a whole new meaning. Not only that, you begin to see much more clearly how and why I integrate geomagnetic storm energies into my research.

Chapter 12

[61] Cyprus, Sheri, and Heather Bailey. "What Is Habituation?" WiseGEEK. March 04, 2018. Accessed March 18, 2018. http://www.wisegeekhealth.com/what-is-habituation.htm.

[62] Piccione, Giuseppe, Simona Marafioti, Claudia Giannetto, Michele Panzera, and Francesco Fazio. "Daily Rhythm of Total Activity Pattern in Domestic Cats (Felis Silvestris Catus) Maintained in Two Different Housing Conditions." *Journal of Veterinary Behavior: Clinical Applications and Research* 8, no. 4 (2013): 189-94. Accessed May 6, 2018. doi:10.1016/j.jveb.2012.09.004, 193.

[63] McCraty, Rollin. "The Energetic Heart: Bioelectromagnetic Communication Within and Between People." HeartMath Institute. 2003. Accessed June 01, 2018, 13. https://www.heartmath.org/research/research-library/energetics/energetic-heart-bioelectromagnetic-communication-within-and-between-people/.

Note: This e-book is available for purchase from the Institute of HeartMath at the listed URL.

[64] Ibid., 13.

[65] Zell, Holly. "Storms from the sun." NASA. July 31, 2013. Accessed June 28, 2018. https://www.nasa.gov/mission_pages/sunearth/news/storms-on-sun.html.

[66] NASA, Science @. "Earth's Magnetosphere." Phys.org. March 26, 2018. Accessed April 03, 2018. https://phys.org/news/2018-03-earth-magnetosphere.html.

[67] Ibid.

[68] Ibid.

[69] "The Kp-index - Wednesday, 11 October 2017." SpaceWeatherLive.com. Accessed April 05, 2018. https://www.spaceweatherlive.com/en/archive/2017/10/11/kp.

Chapter 13

Mind Blowing Encounters Part 2 — 2017

"Success is not the absence of failure; it's the persistence through failure."

— Aisha Tyler

Julie arrived Oct. 11, 2017, for a few days of team research. And what a few days they turned out to be! As luck—or that pixie fate—would have it, she showed up while exciting and wondrous sasquatch experiences were occurring with me. I was thrilled to share these with her.

I took her to spots where I had seen plentiful signs of sasquatches. For instance, she was singularly astounded and deeply stirred to see Sasquatch Highway. Anyway, we wandered around that area for a few hours and figured out that most of the tracks headed toward ponds deep within the wilderness in addition to two lakes a few miles away. Hiking to the ponds would take too long, leaving us in the wilderness after dark, so we chose to drive to the lakes for a brief exploration.

October 11: Scouting and visitors at camp

One of the lakes is small, surrounded totally by wilderness; we discovered we could carry our canoes to it from the road so decided to explore it the next day. The other lake is quite large, with ample public access for boating and fishing, and parking for vehicles. Even so, a good bit of wilderness bordered the shores of this lake, too. Nonetheless, once we decided to focus our attention on the smaller lake, we drove back to camp to settle in for the night.

We prepped for our canoe excursion the following day, cooked dinner, and then sat outside the trailer to enjoy the evening and see what, if anything, might happen. The time was approximately 10:00 p.m. The sky was magnificent; it literally sparkled with billions of shiny, twinkling lights; clearly a welcome invitation to view its vivid feast for the eyes. A few chipmunks were scurrying around the edges of camp, cicadas were chirping lightly, and we heard a single far-off hoot of an owl; maybe it was telling its night prey, "I see you!" Another lovely, serene evening in the wilderness paradise of the Northern Highland geographical area in northern Wisconsin.

Once again, of an immediate sudden, all sound ceased. Complete, absolute silence prevailed in a heartbeat, permeating the entire area with a stillness so deep that it almost hurt our ears. We had been watching the sky thinking we might see the aurora borealis. After glancing all around, our eyes returned to the sky. No color; merely the panoramic spread of a brilliant field of stars.

We fleetingly saw a *falling star*; you know, that brief bright white light showing the visible path of a meteoroid as it crosses the threshold of our atmosphere to become an actual meteor before burning up or crashing somewhere on Earth. Always super cool to see this eruption of light burst across the breadth of the sky in the blink of an eye; we assuredly let it tickle our fancy for a moment or so. As the meteor was moving away from us, we once more turned our attention to the darkness of the shadowy wilderness surrounding our campsite.

With rapt attention paid to the woods, we began hearing sounds suggesting several "somethings" were moving slowly toward camp. It is not uncommon for us to hear one or two somethings around camp, or even three. Hearing more than three is highly unusual. This time, we heard several headed our way—and they were coming from different directions! We glanced at each other in complete wonder. We didn't speak because it was clear to both of us that we were thinking the same thing, or certainly something similar. How many are coming to visit tonight? Holy WOW!

By 10:20 p.m., we had eight visitors! That's right! There were **eight**! Oh, my! They had surrounded our half-circle campsite. We saw none of them because they stayed within the ghostly shadows of darkness permeating our perimeter view. We certainly heard them, though, and were able to pinpoint their physical positions based on their non-stealthy movements. Photo 13.1 shows the half-circle campsite from the very edge of the woods at its perimeter. As you can see, I chose a rather small space in which to create my base camp. Also, you might see where I always parked the truck.

Photo 13.1: Half-circle campsite 2017

To say we were excited is to casually minimize the moment. Julie and I were practically beside ourselves with pirouetting delight over the knowledge that *eight* bipedal anomalies chose to visit us for a look-see at what we were doing. Goodness!

I glanced at Julie, grinned like a silly young boy with absolutely no sense at all, and said, "Hope these suckers are friendly!" Guess it's hard to keep the little boy inside a grown man from peeking out time-to-time, eh? Here he came again! Just so you know; that child was smack-dab tickled silly!

The eight bipedal anomalies were close enough we could hear them shuffling about; however, they remained always just out of view. Darn it! A couple of them uttered vocalization sounds; one sound was low and guttural while the other was a bit higher pitched. We guessed that the low guttural sound was a male and the other a female or youth. This assessment was based on many earlier vocalization experiences.

We had a small campfire burning while we listened to the sounds. From time to time, we set another small piece of wood on the fire. It crackled and sizzled quietly as we sat straining to hear their movements. Obviously, the sasquatches were not alarmed by the fire or us. In turn, we had no reason to be alarmed by their presence. Sure, and we wished we could see them—or at least spot visible movement.

Alas for us! They stayed just beyond the periphery of our vision, teasing us with subtle, haunting movements and an occasional vocalization. Even so, we had surely been visited by **eight** spectral creatures on this auspicious night. We felt exhilarated! We knew there were eight visitors due the obvious footprints left behind. More on these eight visitors a bit later.

I'd guess it was around 11:00 p.m. when we finally extinguished the fire and went inside for the night. We, of course, chatted about our visitors for a few minutes before shutting down and going to bed. We had physical activities planned for the next day and needed a bit of rest, so staying up too late was not a wise choice. See? We learned a lesson and implemented that which we had learned. Smart, eh?

October 12: Canoeing on a lovely lake

Ah, yes! This day we tackled exploration of the small lake. We drove to the location, carried our canoes and equipment to the edge of the lake, loaded up, and paddled out to see what we might see. I was convinced that somewhere along the lakeshore, we would find clear and compelling evidence of bipedal anomalies using the lake for drinking, bathing, or possibly fishing. I was also hoping we'd find castable footprints.

The sun was shining brightly, creating mirages on the tranquil, iridescent water of the thick, densely growing environment surrounding this wilderness beauty. We were at once enchanted; to the point, I might add, that we didn't speak for several minutes. Rather, we paddled slowly, occasionally stopping for moments to take in the stunning picturesque vista.

It was singularly serene, remote, and vibrantly charming. A view well worth our efforts to partake of while appreciating the fullness of nature's ever exploding display of a gorgeous panoramic scene. The Northern Highland geographical area in northern Wisconsin is surely a wilderness of sheer beauty, daily expressing itself as a dynamic portrayal of visual delight to all who still themselves long enough to view it! That included us!

As waves of pictographic exquisiteness swept over us, we continued to paddle silently for a few minutes more. We were near the center of the small lake when we slowly turned our canoes a slow, deliberate 360 degrees to soak in this all-embracing and stunning vision of nature. Upon completion of our hushed turns we looked at each other, smiled in appreciation and gratitude, and then spoke for the first time since embarking from the distant shore.

We chose a direction to go toward the shoreline on the far side of the lake and quietly paddled that way with clear intent intermixed with high hopes. This, while being gently massaged with a tranquil brush of nature's bountiful beauty.

We arrived at the shoreline and split up to reconnoiter in both directions. Within about 20 minutes, we realized that landing access was nil; the shoreline was seriously steep, which prohibited easy landing and alighting from our canoes. We reconvened to discuss our options, and finally decided to stay together to round the lake together. Off we went. Crap! The shoreline along the area we paddled continued to be way too steep for a safe landing. Very disappointing!

I finally saw an area where I thought we might, with a bit of effort, be able to exit our canoes and reach the top of the precipitous shore line. Okay. My sights may have been set a bit high. It took me approximately

12-14 minutes of strenuous effort to get out of my canoe onto the shoreline. Dang! Should have brought an old-fashioned rope ladder with me! My cap took a bath while exiting the canoe, but the rest of me stayed dry. Not only was disembarking not easy, it was a bit treacherous. Nonetheless, I struck shore in the end. After resting a few minutes and assuring Julie that I was alright, I began looking around.

Such remarkable beauty! Nature sure does know how to impress this ole' boy. As I wandered around on shore under a thick umbrella of closely growing spruce trees, Julie continued to scan for easier access along the shoreline. While she did not achieve success, I certainly found the evidence I expected; bipedal anomaly signs were everywhere! Goodness! There were footprints galore, along with signs of bedding near the lake. None were castable because of the deep pine needle bedding under the thickly thriving spruces. Even so, the area was rife with signs, which served to pique my interest even more.

I wondered for a moment how sasquatches could use such a treacherous location to enter the water. Then I smiled to myself as I recalled sasquatches are nearly twice my size and most likely at least quadruple my strength, if not more so. This location would not be an issue for them; only for puerile humans such as me! Hmm!

Anyway, after wandering about for a quarter-hour, I decided that the shoreline did indeed produce evidence as I had thought it would. The next step was to find a spot for *easy* and *safe* access. It only took me about half the time to reenter my canoe as it did for me to disembark. I saw Julie still slowly checking the shoreline for an accessible ingress/egress spot and canoed to her. We updated each other and decided we would continue wandering the perimeter of the entire lake to find a spot we could use as a suitable landing.

Off we went around the tranquil, shimmering, mirage-like surface of the lake. We were truly blessed this day with a combination of perfect weather, sound water conveyances, and a stunning, sweeping scene from nature. It is also true that we felt a quiet comfort knowing we had each other for back-up, as well as being able to share the striking and exquisite beauty of the scenery. Sharing a day like this with a dear friend is unusual, yet deeply touching.

Embarrassing Story

I had no intention of telling this story; however, Julie thought some of the readers might appreciate it. By the fall of 2016, I decided kayaks would no longer work for our waterway trips. Remember the *three hard lessons* of 2015? That trip showed me we needed different water conveyances. We ordered handmade canoes from Hornbeck boats in Olmstedville, New York.

I ordered a 12-foot canoe and Julie ordered a 10-foot one. The canoes were designed to fit our bodies, and we had each tested them and set adjustable foot pads to our needs. The day Julie and I explored the small lake, she and I had carried the canoes to the lake shore and set our equipment on the ground. We walked around bit, talked a bit, fussed with our equipment, and finally loaded the canoes and hit the water paddling. I don't recall which of us launched first; even so, I noticed Julie adjusting her foot pedals and wondered why she did so inasmuch as both of us had tested our canoes prior to this and we each had the foot pedals set to our liking. Something wasn't right!

We eventually figured out that we were in the other's canoe! Say what? Talk about feeling silly while trying to be professional, adult, and serious! See? Sometimes we become enamored with the task and forget common sense! Anyway, after trading canoes, we had the wonderful day described above. I've tried unsuccessfully to put this episode behind me. Sadly, Julie still giggles about it from time to time. Okay … I may be smiling a bit right now. Go figure!

We slowly canoed around the entire lake. The only spots we found for easy landing were two: the area from which we launched and another one a few hundred yards from it. The second one came complete with a small overnight camping area; however, as this one was not far from our original launch site, it really wasn't what we were looking for. This, because we would have had to spend several hours hiking around the lake through the forest and another several hours hiking back. We weren't prepared for that, so it didn't serve our purpose at the time. We might go back some day, set up an overnight camp, and take the daylong hike. Who knows?

Not far from these two locations, we found a flat, seriously swampy area. I decided to pull myself into a spot that I hoped held a proper landing zone. It didn't. In fact, I found the effort was not worth the trouble because I nearly got myself so stuck in muck that I thought I would have to exit the canoe and slosh through the nasty muck while dragging the canoe to withdraw from the area. I was happy Julie couldn't see my dilemma; this was awfully embarrassing. I finally managed to pull my way free by moving tediously backward with the canoe. As I neared the spot where I could begin paddling, I looked up and noticed Julie sitting nonchalantly in her gently rocking canoe with her feet up, resting, drinking water, and grinning at me.

Damn! Busted for sure. Oh well, not my most embarrassing moment in life, although she may have commented about wondering if she would need to rescue me. Gotta love dear friends—even if they do have a droll sense of humor.

Regardless, by this time, the day was waning, the sun was beginning to set, and we needed to get off the lake and back to camp. We returned to our starting point, loaded the truck, took off and arrived at camp well before dark.

We had a specific spot near camp where we left a few apples as gifts each evening for the wandering sasquatches who regularly visited us. After placing the apples, we ate dinner and once again sat outside to see if anything would happen.

Our woods friends visited us again, just as they had the night before. Once again, we heard subtle movements and minor vocalizations; they were not deterred by our campfire or us moving about and discussing their visitation. A real difference this night was the way Julie and I constantly moved about the campsite, stepping in and out of the trailer, and conversing in normal tones. We retired around 11:00 p.m., wondering what the following day would bring. Time would tell.

October 13: Julie's final day of research for 2017

This was Julie's last full day of the 2017 research trip. We didn't do much in the way of research today. She began packing and readying her vehicle for the trip home. We did drive to town to pick up apples and have breakfast at a local restaurant that morning. Other than that, we wandered the forest while staying close to camp.

At one point, we decided to set up a trail cam that evening, a highly unusual choice for us. I placed it high in a tree overlooking the area where we left apples each evening. We really didn't expect to capture anything with the camera trap; however, the temptation to try was simply greater than our ability to resist. Wonder if this suggests our internal children momentarily clawed their way out long enough to affect our decision? Youngsters!

We stayed up late this night, listening to our visitors move about in the ethereal woods. From time to time one or more of them uttered vocalizations, none of which sounded alarming or otherwise inappropriate. Rather, they seemed to be prattling on about the silly humans wandering around the campsite almost as if they enjoyed our momentary madness. Indeed, we acted and moved about as if their presence was a natural event in our lives, and at times, giggled like children risking an unknown danger. Yes, we enjoyed every minute of it. Hmm! The children within sneaking out to play, or a simple spate of unexpected silliness? Frankly, it mattered not a whit to either of us that night! One thing for sure; we gave those suckers a smile or two!

Fun! Research is not all boring! Sometimes we must simply placate the still yearning child within! Yup!

It was around 1:00 a.m. when we finally called a halt to the playful evening and went to bed. I sleep in the front of the trailer and Julie in the back. At about 1:20 a.m., Julie quietly asked me if I was rocking the trailer. Nope! It wasn't me. Dang! The trailer gently rocked for several minutes. I believe this was the first time Julie had felt our visitors lightly rock the trailer. Quite an exhilarating feeling, I say! And I was immensely pleased that Julie finally had an opportunity to undergo this dynamic experience. Maggie and I have experienced this several times, and now Julie had also felt the trailer rocking. Neat! A fitting end for her final night in camp, eh?

I lay awake late into the night, or well into the morning if you prefer. I couldn't stop thinking about all that had happened during this eventful trip. Also, I knew Julie was leaving for home shortly after we got up the next morning. We had delighted in an action-packed few days together and I was sad to see it end.

I didn't plan to leave for a few more days yet didn't know if anything else exciting could even happen. Besides, I was feeling fatigued and knew that I needed rest before I could go out on my own again. I finally fell asleep wondering if the trailer would rock once more during the rest of the night. It didn't—or, if it did, I slept through it.

October 14: Wood-knocking symphony?

We both arose later than usual this day; I heard Julie get up and go outside. I was so tired that I didn't wonder if she was loading her vehicle readying for departure or just sitting alone and thinking about all we'd experienced. Neither did I check the time; I simply closed my eyes and fell back asleep.

It felt like only moments later when I heard the trailer door reopen and her say something like, "Roger, you've got to come outside … right away!" She closed the door without stepping fully inside. I had noted a bit of tension in her voice, so I jumped out of bed, grabbed footwear and a jacket, and went out to see what was up.

I found her standing at the rear of the trailer; she motioned me to join her. As I approached her, I heard a knocking sound. I quickly glanced all around and noticed that two local hunters had parked their vehicle in its usual place not far from our camp to hike into the wilderness while hunting. I saw nothing else around us.

When I initially joined Julie, the knocking sound was being made by a single entity—a human or bipedal anomaly. Within a couple of minutes, a second knocking sound merged with the first. We looked at each other with questioning eyes. A couple minutes later, we distinguished a total of *five* separate knocking sounds. Shortly thereafter, we counted a total of **eight** distinct knocking sounds. Whatever was being used to produce the sounds must have been from several types of wood in varying lengths or, perhaps, differing diameters of wood or different sized rocks, because each of the eight sounds were distinct, different in tone, and readily recognized as such. What in the world?

We briefly glanced at each other again, this time with a combination of disbelief and amazement. Another few moments of knocking and then it began to sound like a wood knocking symphony! Various tones, pitches, and strength of sound emanated outward from the location of the knocking. We were flabbergasted, and nearly beside ourselves with a mixture of awe and a sense of unbelievable joy that we were being given such an implausible gift of wilderness sound! Truly.

We had just been blessed with a wilderness symphony the likes of which we had never heard. Imagine the joy we felt as we assimilated this experience. I'd almost swear we could feel the sound waves momentarily embracing us as they swept past us traveling to who knows where. Surreal!

Julie had marked the time of the first knocking sound and then when the last sounds faded, she marked the ending of them. The wood knocking symphony had lasted for a bit over 17 minutes! That's right! A **17+ full minutes** of a wood knocking symphony as we stood near the end of the trailer! Startling, to say the least. Staggering, to say more. Unless one looks at the subtitle of this book! 17+? Hmm!

As we turned to head back to the trailer door, I noticed that the hunters had left. Apparently, they didn't take the same view of the wood knocking as we did. Belief ruled that experience and the hunters perceived and reacted according to their belief systems—as did we. The hunters fled … we were blessed. To be clear, some folks within the sasquatch community believe that wood knocking is a signal to get the heck out of Dodge! I suspect that the hunters left because of this belief. Clearly, Julie and I felt differently.

To me, the wood knocking symphony evoked a similar feeling of amazement, along with a sense of being blessed with such exotic sound waves. Quite the extraordinary experience, I say! And certainly, one that I am not ever likely to forget. I seriously doubt that Julie will ever forget either. Anyway, upon reentering the trailer, we cleaned up, made coffee, and prepared for the day. We took our coffee outside to discuss the wood knocking concert. As we sat down, we looked up and noticed what appeared to be fingerprints on the inside of the front flap of the trailer awning.

Memories of an Earlier Symphony

This wood knocking symphony put me in mind of a time years ago when Maggie and I were visiting the Chichen Itza pyramid in Mexico. Chichen Itza was a large pre-Columbian city built by the Mayan people of the Terminal Classic period. It is a well-known archaeological site in Tinúm Municipality, Yucatán State, Mexico. Maggie and I were meditating at the base of the pyramid.

We used the Mayan form for our meditation. A few minutes into our meditation, we felt and heard several folks move up behind us. Shortly thereafter, we heard five distinct clapping sounds. After a few minutes the clapping sounds came faster and louder until they reached a nearly deafening crescendo of rhythm—and then immediately stopped.

Unbelievably, the sound seemed to rise to the top of the pyramid, enter a sound chamber atop it, and then return to us, ultimately bathing us in an inexplicable groundswell of exotic sound waves. We learned afterward that a group of Mayans had done this for us because they saw us honoring the Mayan meditation form.

Sadly, it was illegal for Mayans to perform a Mayan meditation on the pyramid property. It turned out that they wère beholden to us for doing so in their honor. Truly a blessing that remains with us to this date! Another memory from years gone by that will never leave the two of us.

Following a couple sips of coffee, I snapped photos of the prints. We noticed only four fingers, no thumb. Odd. Also, these prints looked a lot like the dusty prints I'd photographed on the tailgate of the truck shortly after I first arrived at camp. Photos 13.2 and 13.3 show those prints.

A slight tear in the awning appears just above the first finger in each photo. The tear was not there before, for I always check the awning before each trip to make needed repairs.

Photo 13.2: Dusty fingerprints on inside of awning

Photo 13.3: Dusty fingerprints; slight tear above the 1st finger

If I had to guess, I'd say that the fingerprints are from a left hand. Looking closely to the right of the 1st finger, it appears that a thumb might have been placed there. Whatever, one more substantial sign that sasquatches are visiting the camp and feel comfortable being around us. Sweet!

Okay. From a wood knocking concert, to hunters scampering to their vehicle and leaving during the wood knocking concert, to seeing visible dusty fingerprints on the inside of the trailer awning flap, to the trailer rocking the night before, to… Julie was truly loath to leave. Yet, work responsibilities demanded her presence at home. Bummer.

We checked the trail cam disk for photos. Just what you'd expect; not much, other than a few strange lights and a blob or two. We returned to our coffee and chatted excitedly about the wood knocking concert. We had not tried to pinpoint their location; we were so astonished we didn't bother with important details such as finding the location from whence the amazing sounds from their symphony originated. However, I was convinced they were no farther than 75 yards from the backside of the trailer.

Following Julie's departure, I walked out to see if I could find signs to show where they stood during the concert. Sure enough, they were about 70 yards from us at the time of the symphony. That's 30 yards less than the length of a football field! They had stood in a relatively small half-circle facing our direction while performing for us. And, remember, this incident happened between 8:00-8:17+ a.m. Highly uncommon! Our typical encounters with sasquatches almost always occur just prior to or after dark, which makes this experience a remarkable one.

An utterly unusual and astoundingly unforeseen several days of exciting and dynamic experiences! Might be a bit of a challenge to digest and acclimate to such startling realizations of an alleged altered reality. Hmm! Altered reality? I think not … at least, not to us. The experiences were as real to us as is each breath we take. Cognitive dissonance? Nah! Unbounded joy!

I spent the rest of the day lounging around camp, although I did revisit the spot where the sasquatches entertained us simply because I needed to be sure I hadn't misread the signs during my earlier excursion.

I hadn't. Photo 13.4 shows a half-circle on a slight hill; the eight sasquatches stood there while performing the wood knocking symphony.

Photo 13.4: Eight sasquatches stood while wood knocking

That evening, my bipedal anomaly friends returned in force. All **eight** of them! This time, though, those suckers came close enough that I detected vague shadows. Guess what? Unless my eyes deceived me, those creatures are frickin' **huge**! I chatted with them for a couple of hours. They refused to come close enough for me to see them clearly; however, three of them did, in fact, chat back at me. Of course, I have no clue what they were communicating, any more than they likely understood me. Still, we had a nice late evening chat; a memorable start, eh? I was thrilled they had returned enmasse.

I decided to try something different this night. I have a small table set up near the firepit; it sits 10 feet from the trailer door. I had two bunches of green grapes in the fridge. I grabbed them and placed them on the table. I then took a grape from the bunch on my left, held it up high, pointed at them, pointed at me, ate it, rubbed my stomach, and said, "Yum." Next, I picked each bunch up separately, held it up high, pointed at me, pointed at them, moved my hand from the heart area of my chest to them, and then laid it back on the table.

I performed that routine several times throughout the evening before going to bed. I kept hoping one of them might step out to take a bunch before I went to bed; none did. Considering the late-night Julie and I had the night before, I was too tired to stay up late this night. I went to bed around 10:00 p.m. I fell quickly asleep and didn't awaken until 7:00 a.m. the following morning. After making coffee, I stepped outside and much to my surprise, the bunch of grapes on the *right* side of the table were gone. The bunch on the *left* side were still there.

Hmm! I wonder if they thought I was offering only the bunch on the right because I had eaten one from the bunch on the left? Or, maybe they're germophobes? Once again, I'll likely never know. Even so, this was yet another delightful experience added to the whole of my research. See what I meant earlier? Piece by piece, layer by layer, this research is taking me on a journey of observational understanding of sasquatches.

Regardless, habituation and familiarization assuredly played out in real time throughout the entire 2017 extended trip. I was amazed and thrilled. Unbeknownst to me at the time, though, the most exhilarating experience with them was yet to come. Wait until you read what happened the following night! My … my … my…

October 15: Bountiful surprises!

I drove to town this day for breakfast. It was another lovely day in the Northern Highland geographical area, with mostly sunny skies and no threat of rain or other adverse weather. Leaves were falling because the season of fall color was concluding. Those still on the trees softly fluttered as a steady light breeze teased them into a flickering dance that stirred a visual sensation of nature's graphic artistry. Stunned with the visual beauty, I missed the message those dancing leaves were sending—

telling me that the eight sasquatches would return to camp in a totally unexpected way!

After leaving town, I decided to check out my so-called Sasquatch Highway. I parked the truck in its usual spot and walked into the woods, carrying only water and a few crackers because this was merely an eyeball check and not an extended hike. Well, I also had my compass and a few other typical necessities; after all, this is, of course, a wilderness.

I snapped a few photos around the area of Sasquatch Highway and then climbed a bit of a hill. Walking the ridge allowed me to see down the hill and into a narrow valley which Julie and I had checked out earlier. The weather was perfect, and I walked slowly, enjoying the breathtaking exquisiteness of the fall forest.

Wait! Pretty sure I had spotted movement, I stopped to focus on the area from which I thought I had seen something move. Shoot! I either imagined it or, perhaps, a bird had flown along the edge of the forest. Not ready to call it quits, I stayed still for about 10 minutes while continuing to focus on the area.

Sure enough! A sasquatch was standing just at the edge of the forest on the other side of the valley trail. I couldn't make out much detail. It was, however, standing still and looking directly at me. Hmm! We seemed to be intently peering at each other, and then it moved just enough for me to get a clearer view.

Oh, my! Another female! She stood just at the borderline between the forest and the valley trail. How dizzyingly exciting! This was my **second** viewing of a female sasquatch this trip, and my **seventh** view of a live sasquatch since the fall of 1999 when this unusual story began.

I estimated she was about 120 yards from me; as such, I am not comfortable suggesting that I can offer detail on her. The photo I snapped of her is an absolute blob. I did not have my zoom camera with me, and my cell phone camera did not pick her out from the forest background. Oh, well! At least I tried for a photo this time!

Strangely, she stood still while I snapped the photo. Maybe she knew it wouldn't take? Whatever! After that, we watched each other for almost 60 seconds before she coolly stepped forward one step, took a final peek at me, then turned and moseyed out of sight into the forest. Sweet! This trip had certainly presented me with astonishing experiences. After watching the empty forest for a few more minutes, I turned, hiked back to the truck and drove to camp—with a contented and happy heart!

Arriving back at camp, I began preparations for the evening's sitting. Because the camera trap in the tree bore no fruit in the form of photos or videos, I decided to place it inside the trailer pointing out a window toward the woods at the edge of the half-circle campsite. I know! I know! The glass in the window was bound to cause reflective issues, and it certainly did so.

All the same, I thought that maybe my familiar bipedal anomaly friends might not mind it as much if it were indoors. Silly me! Also, remember this camera was one of the new "no glow" ones, so maybe an intermittent infrared radiation light wouldn't bother them if seen through the window. Into the window it went, and I used a blanket to block off the back of it to prevent reflections in the window from the lights and activity in the trailer.

Once the camera was set to my satisfaction, I started a small campfire. With the fire going, I decided to cook dinner over it. After all, it's hard to beat a campfire cooked dinner while deep in the wilderness, eh?

I intended to complete my chores before the sun set. I cleaned up after dinner and split just enough wood to keep the campfire sizzling late into the night. It was beginning to get chilly, so I changed into comfortable and warm lounging clothing and had a jacket and gloves handy for later.

The final chore on my agenda was to place a few apples in the familiar spot. I grabbed eight apples and headed over to the apple spot, which lay just beyond a few pines bordering the northeast edge of the campsite.

We could not see the apple spot from the trailer; however, we could hear any bipedal anomalies when they stopped to partake of the apples. So, after delivering apples, Maggie, Julie, and I would always sit quietly just after dark listening for visitors to walk in and enjoy the gifts. Photo 13.5 shows the pines at the northeast edge of camp behind which was the apple spot.

Photo 13.5: Apple spot behind the pines — left side of photo

I walked past the truck—always parked right in front of the trailer tongue—and headed to the apple spot. I paid no attention to my surroundings, being totally focused on prepping for the night's possible adventures. About half-way to the apple spot, I heard a mid-pitched somewhat guttural vocalization. Astonished, I stopped and slowly turned my head.

Holy Scat! I saw a bipedal anomaly standing approximately 25 feet from me! 25 feet, I say!

It was a female and the mid-pitched vocalization had come from her. I stammered in bewilderment, "Well … hello." It looked as if she were headed to the apple spot. She immediately turned and began to casually walk parallel to me in the opposite direction from where I was heading. She repeatedly turned her head and shoulders to look at me and twice more made the same vocalization.

What is most fascinating about this occurrence is the time: it was only *5:15 p.m.* and the sun had **not** yet set—and wouldn't until around 6:12 p.m.—nearly an hour later! Really? The vocalization she continued to utter while slowly walking away sounded an awful lot like the "Well, hello" I shared with her. My heart flutters as I write this, and I feel a similar adrenaline flowing freely throughout my body now as I did then. Further, I'm as mystified today as I was at that moment.

I've no clue why this friendly sasquatch showed herself to me in broad daylight. Not only that, she did **not** hurry away from me. It was almost as if she knew I was delivering apples to the apple spot and wanted me to get it done so she and the others could partake of the early evening treat. Geez! It seems sometimes as if all I have around me are bosses and more bosses! Does it never end?

Even more unsettling, as she walked away from me, it appeared as if her feet and lower legs were disappearing into thin air! No frickin' way!

Yet, it happened. I saw her; I heard her; I watched her walking casually away from me as she meandered toward the woods; I watched her feet and lower legs seem to disappear into ... nothingness! To answer an obvious question, I wasn't drinking, and I do not engage in recreational drugs. Although, at the time this incident occurred, I may have thought I should damn well start! This stuff is enough to scramble anyone's brain waves. Neurons gone wild? Hmm! Cognitive dissonance? Maybe. More on this idea a bit later in the chapter.

At the time of the incident—5:15 p.m.—the temperature was 45 degrees F, the dew point was 36 degrees F, the humidity was 72 percent, the barometer was steady at 28.4 inches, and it was partly cloudy, with about a 5-mile-per-hour light breeze blowing from the west-northwest. Additionally, the Kp index was 5 and the Bz component of the interplanetary magnetic field was south, which means we were under a minor geomagnetic storm. Further, this followed a stretch where we had undergone a total of **18 days** during which Earth experienced geomagnetic storms or the magnetosphere was otherwise unsettled.

Basically, Earth's energy interacting with atmospheric energy, combined with solar energy, were such that a possible "induced transparency" was highly probable. Call me crazy if you wish; however, I honestly believe I watched such an event unfold as I stood there speechless, watching as her feet and legs seemed to disappear into thin air! Whew!

As difficult as this is to believe, though, I'm relaying the incident to you precisely as the event unfolded. Even more beyond belief, once she stepped into the woods, I walked over to check for tracks. I followed her tracks to the point her feet and legs began vanishing and sure enough—she stopped leaving tracks. Okay. Now what? Bring on cognitive dissonance! Is sanity as fleetingly silly as it felt at that moment? What truly is reality? Anyway, this was the **eighth** sasquatch I'd seen since 1999 and, assuredly, the closest I'd ever been to one!

I'm reminded of a simple, albeit, profound, distinction between two great thinkers, Plato and Aristotle. As I recall from studies ages ago, Plato believed that reality essentially occurs within the mind, while Aristotle suggested that reality is tangible. I sort of wish that both those fellows were here right now to aid me in discerning if that which I saw and experienced as physical, is a construct of my mind, tangible, or a bit of both! Whatever! Those dudes are both dead and if they were here now to advise me, I'd be in a greater funk of confusion than the seeing of a sasquatch's feet and legs disappear gave me. Shiver me timbers!

Returning to my then reality, whatever we choose to name it, I recorded footprint size, step, and stride. The disappearing Queen's footprints measured $21\frac{1}{2}$ inches long, $5\frac{7}{8}$ inches at the heel and $5\frac{1}{8}$ inches at the midpoint. I measured her average step length at 4 feet 8 inches and her average stride at 5 feet 6 inches. I measured her step length from the toe of her right foot to the toe of her left foot in the track, and her stride from the toe of her right foot to the heel of her next right footprint. I took nine step measurements and divided for an average; I took eight stride measurements and divided for an average. These are typical calculations for step and stride measurements. Refer to diagram 7.1, Chapter 7 for details on step and stride measurements. Next up, I offer a description of the female sasquatch I saw that was standing a mere 25 feet from me when I first saw her.

Description of Female Sasquatch

The female's body hair was about 4 inches long. It covered her from head to foot, although it appeared as if the hair on her feet and the backs of her hands was much shorter than her body hair. The hair was predominantly medium brown. When she turned to walk away from me, I noticed two narrow vertical lines of hair on her back that seemed to be a bit lighter. I did not see hair on her face and her head was more rounded than pointed, although I easily distinguished a slight conical point at the backside crown of her head. The hair on her body seemed clean and groomed, not matted, twisted, disheveled, or otherwise unkempt. I'm not suggesting that she carried a comb and brush with her, merely that she looked clean and groomed.

Although I initially believed she had no neck, she turned toward me as she walked away, and I saw that she did indeed have one. What confused me at first was the incredible amount of strength in her neck, making it look as if she lacked one entirely. While I can't say that her neck was identical to human necks, she certainly had one. I do not, however, believe she could move or flex her neck as readily as do humans. Curious.

Her eyes were symmetrical above her nose and quite large; her nose was close to her face without being completely pug-nosed, and I could clearly see her nostrils. The nose was black, and her eyes a deep brown. She also had symmetrical and somewhat bushy eyebrows above her eyes. The hair on her head was a bit or so longer than the hair on her body, much like you'd see uneven near shoulder-length hair on a human. I saw no hair on what I perceived as a leather-like face. The leather-like face was not black, as I thought it might be. Rather, it was a dark brown hue, much like her eyes and very much like tanned leather. I noticed slight wrinkles appear in the cheeks of her leather-like face when she pursed her lips or slightly opened her mouth for a vocalization. This seemed very humanlike to me.

Her shoulders were axe-handle broad, and her arms hung down with her fingertips hanging just at her knees. I did not see hair on her palms; of course, I only had a few glimpses at them as she was walking away from me. Her arms were certainly muscular, evenly proportioned with muscles along both her forearms and upper arms. I decided she was female because I noticed breasts, although they were not ponderous like some descriptions I've read. They moved naturally with her body much as do breasts on a woman who walks unclothed.

I estimated her height between 7 feet 8 inches and 7 feet 11 inches. Her weight was difficult to discern due to her symmetrical and evenly toned body; even so, I'd guess her weight at around 750-775 pounds. She was a striking figure, and even though her thighs were large and muscular, and her calf muscles thickly corded, she did not look at all like the animal many pictures I've seen of sasquatches suggest them to be.

She unquestionably was as close to human looking as she was to looking like a creature, although she was obviously *much* larger and way stronger than humans. Her buttocks did not protrude out abnormally; rather, her derriere appeared evenly proportioned.

Her feet were, of course, large—and, they would need to be to handle the weight this sasquatch packed around. I was not able to discern specifics about her feet, although I did count five toes on each foot. Geez! The dang toes were, like, seriously large. Her legs did not move like human legs do while walking; rather, her knee joints appeared to move differently humans do. Even so, her movement seemed far smoother and more efficient than humans.

The overall symmetry of her body build caused me to believe she was a human hybrid of sorts. Or, perhaps, a form of pre-human? I simply do not know enough to adequately categorize her. Although some species of sasquatches may, in fact, look like or relate to, one of the ape genera, this sasquatch definitely did not fit an ape genera model. There was no evidence of a tail, and both her feet and hands looked mostly human, although obviously much larger. Her fingers were long and symmetrical to the palm, much as are human hands and fingers. Her thumb appeared to be placed somewhat differently than on a human. It seemed almost as if the thumb were a bit closer to her wrists than human thumbs. Although I detected movement in her fingers, I did not see the thumb move.

When she uttered vocalizations, her lips opened as some of the sounds came from between her teeth and I saw proportioned, evenly-spaced teeth; I did not see fangs. Her lips looked similar to human lips, albeit a goodly bit larger of course, and longer to accommodate the long mouth on her large face; her lips were a lighter brown than her darker, leather-like face. When she opened her lips to utter a vocalization, I saw brief contractions and expansions of her cheeks along each side of her nose. Again, this seemed like something we associate with humans as they speak.

One final observation: she rarely blinked her eyelids. Near as I could tell, she blinked only twice during our entire encounter. She surely did not blink as often as most humans do. Her eyelids seemed puffed out more than on a human and, her eyelashes were quite long. I believe, though, that her eyelashes were balanced with her face; that is, her eyelashes on a human would be exceptionally long, while, on her, they fit uniformly to her face. Due to the near shoulder length head hair, I did not see much of her ears. The tops of her outer ears appeared rounded like humans; I could not tell if she had earlobes.

I'm still not sure what to say about this experience. Guess I'm still processing it, and likely will be for the next decade or so!

As I write this, I can see her in my mind's eye. I detected what I believe was both intellect and curiosity expressed in her deep brown eyes each time she looked at me. She seemed completely capable of expressing emotion through her eyes and vocalizations, yet I did not sense fear, anger, hatred, or angst of any kind. Instead, she seemed as curious about me as I was her, although she was clearly not comfortable staying within 25 feet of me. Indubitably, that was okay with me at the time! I will truly not ever forget the full vision of this sasquatch. An absolute unbelievable moment frozen in time!

And this, folks, was merely the beginning of a literal cherished—albeit, surreal—night! Truly!

The Evening Continued

After she disappeared, I finished placing eight apples at the apple spot and straightened things up around camp. At that point, I decided I was ready for whatever might happen next—or so I hoped. As an afterthought, I made coffee to help keep me warm while sitting outside because the temperature was quickly headed below freezing this night. Not only that, but camp coffee is truly unequaled. Take that, *Starbucks*!

I settled in outside at around 6:15 p.m. It was nearing dusk and was quiet. None of the typical early nighttime sounds could be heard. Actually, a deep, probing silence pervaded the area as all the typical night creatures seemed to hide from these gentle giants. I knew they were near because I heard movement noises over by the apple spot and in the pines directly behind it (refer to photo 13.5).

While sipping coffee, I heard them moving from the apple spot into the wooded area surrounding the half-circle campsite. My heart started beating faster with an anticipation not experienced on other nights. This most likely occurred because I had earlier seen a female sasquatch **25 feet from me**. Regardless, I was so excited I was tingling all over and thought I might pee myself.

When the bipedal anomalies stopped moving around, they were positioned just as they had been on earlier nights. Eight of them were spaced around the half-circle campsite, although there was a major difference with them tonight. They were close enough to the edge of the shadowy woods I made out their shapes! Damn! These beings—whatever they end up being called—really are massive! I mean, they're big enough to scare the scare out of scare! If that's even possible.

Obviously, my heart continued to flutter for a while as I sat watching them watch me. I'm totally convinced that they were as curious about the strange human who had been tramping around their territory as I am them. They were getting as close a look at me as I was them, although I suspect they have night vision of a sort and were able to see me much clearer than I saw them.

I repeatedly looked at each of them in turn as they stood at the very edge of the half-circle watching me back. I could see only the top of the head of one of them. That one appeared to be kneeling in the pines not far from the apple spot I mentioned earlier. I decided there were three adult females, two adult males, and three youth. I figured the kneeling one was a youth because its head gave me the impression that it, although large, was smaller than the heads of the adults.

After what seemed an interminable amount of time, the largest one spat out several sounds. Clearly, the tones were directed at me. I returned its direct look and responded quite wisely, "Huh?" Goes to show us; we never know how we will react in a situation until we are in it. Again, cognitive dissonance? Maybe.

Okay. Following that brilliant verbal opening, I settled down enough to engage with them the only way I could; I spoke with them as if we were new friends using different languages. Two of the three adult females and the largest adult male conversed with me. While the youth sasquatches left communication up to the adults, they never once acted bored; nor did they fidget around like young children sitting through a lengthy sermon in church are often prone to do. Rather, they stayed attentive to the conversations occurring between the adults and me—almost as if they were studying and learning from the activity. I'd almost swear the youth were focused on our body language and facial expressions as much as they were on the verbal communication.

I used both verbal and crude hand signals while trying to communicate with them. They, in turn, also used crude hand signals and a variety of vocalizations while trying to communicate with me. Most astounding and electrifying experience ever! Wow!

Think about this for a moment or so. I'm sitting in the dark in the middle of a wilderness with no light other than stars and a small campfire. I'm viewing eight so-called sasquatches standing close enough to me we that see each other *almost* clearly. They and I engaged in nonsensical communication using both verbal and hand signals. Neither they nor I really understood what the other was saying. Delusional apparitions … or concrete reality? This experience was as real to me as me typing these words; ergo, my answer is, "The experience was concretized in my reality at the time it occurred." Wanna debate it? Not with me.

Be that as it may, this was certainly an extraordinary, if not a far-fetched, moment. Historic, to say the least. Sasquatches trying to communicate with a human while a human was trying to communicate with them! A literal conversation of sorts occurring between a human and eight sasquatches or so-called bipedal anomalies. Oh, my.

Unheard of, you say? I was there. Impossible, you say? I was there. I willingly partook in the incoherent communication. Three of eight sasquatches willingly engaged with me in return. Three of eight, I say!

Now are you beginning to understand the title of Chapters 12-13: *2017's Mind Blowing Encounters*? I am.

With all that happened thus far, the night was far from over! Indeed! For instance, I had placed a small table next to the campfire to set things on while cooking or tending a fire. The table was 10 feet from the trailer door. The largest bipedal anomaly was standing 52 feet from me at location number seven (see the next section for photo depictions of sasquatch locations 1-8 as they appeared to me around the half-circle surrounding my campsite). This mammoth guy communicated with me much more than the others. Clearly, this one oversaw the group of eight.

Because it was in charge, I decided to try a different, more direct approach of communication. I stepped into the trailer and grabbed my last apple. After stepping back out, I held the apple above my head hoping the largest sasquatch could see it … I'm sure it did. I then moved it to my mouth to pretend I took a bite. I held it back up and lightly rubbed my tummy, making "Umm … good" sounds. I repeated this act several times. Finally, I placed the apple on the center of the small table near the fire pit, repeatedly pointed between the largest sasquatch and the apple while saying several times, "For you … for you … for you." I then touched my heart area and said to him, "From my heart to yours." I repeated this phrase and action many times until I went to bed. Silly? Perhaps … until the next morning when I saw a distorted trail cam video of the boss sasquatch reaching out and plucking the apple from the table! Silly? Perhaps not. Perhaps simply another step toward conscious, intentional interactions with sasquatches.

All eight sasquatches watched this act carefully, glancing back and forth between me and the largest one. The deference seven of them paid the largest one confirmed to me he was assuredly the boss. The boss of eight watched me with studied intent while I performed the silly apple act. Twice he glanced to his left at the female standing in location six. I heard her say something to him and he responded. Following this brief exchange, he returned his gaze to me. My assessment at the time was that he appeared somewhat puzzled, whether with or at what she said is unknown. I believe she was likely the female I'd seen earlier (maybe, the boss of the boss?). Nonetheless, I continued to point at the apple and then at him, reminding him the apple was for him if he wished.

I was hoping he might step out to retrieve the apple while I was standing there. He didn't. To learn more about what happened to the apple, continue reading into the next section, which describes my activities of the day following this extraordinarily eventful-packed night. I stayed outside for about another thirty minutes or so. At 2:30 a.m., I extinguished the campfire and went to bed. Tired is tired, irrespective of the amount of excitement one feels. Actually, exhaustion was more like it.

October 16: The day after

I arose the next morning at 8:00 a.m. needing to document the previous night's experiences. I measured location distances from me as I stood near the trailer door to the eight sasquatches I had seen. I also snapped photos of the locations and what they saw as they viewed me the night before. Photos 13.6 through 13.19 show where the eight sasquatches were placed around the campsite.

Photo 13.6: Sasquatch location 1 — 24 feet away

Photo 13.7: Sasquatch 1 sees me — 24 feet away

Photo 13.8: Sasquatch location 2 — standing

Photo 13.9: Sasquatch 2 sees me — 25 feet away

Photo 13.10: Sasquatch locations 3-5 (from right-to-left)

Photo 13.11: Sasquatch 3 sees me — 37 feet away

Photo 13.12: Sasquatch 4 sees me — 37 feet away

Photo 13.13: Sasquatch 5 sees me — 41 feet away

Photo 13.14: Sasquatch 6 location — standing

Sasquatch six moved from the arrow to the left of the tree to the arrow to the right of tree shortly after they arrived.

Photo 13.15: Sasquatch 6 sees me — 34 feet away

Photo 13.16: Sasquatch 7 location — standing

Photo 13.17: Sasquatch 7 sees me — 52 feet away

Photo 13.18: Sasquatch 8 location — lying down

Photo 13.19: Sasquatch 8 sees me — 49 feet 6 inches away

Considering the marks in the pine needles from where sasquatch eight saw me, I decided this one must have been lying down, not kneeling. It would raise its head from time-to-time, so it could see me, and that's when I saw it. In the end, I believe this one had a physical deformity and was likely carried to the campsite and placed on the ground under the pine trees not far from the apple spot, yet close to the boss. Hmm! A group of sasquatches carry a physically challenged young one a long distance to interact with an oddball human. Just keeps getting more thought-provoking, eh?

October 17: A 'Blob' Photo of Sasquatch Position Seven

Okay. Now we get to the weird stuff! Say what? You thought the earlier experiences were weird? It's all a matter of perspective, I guess.

I went through the photos and videos taken by the camera trap the night before and even early that morning. All I can say is the ones I reproduce here fall firmly under the heading of *blob* photos, with a bit of induced transparency thrown in for good measure. With that said, I sincerely hope you see what I see when I view them. They are indeed quite remarkable.

Photo 13.20 shows the top of a head and eyes of sasquatch number seven, the largest of the eight and the one obviously in charge. I grabbed this snapshot from a short video taken by the camera trap placed in the window of the trailer. The eyes are looking directly at the camera in the trailer window; the time stamp is 6:45 a.m.

Photo 13.20: Sasquatch face; eyes looking at camera in window

Also, please check out the pine tree just to the right of sasquatch seven. There are several natural white spots on the tree in a vertical line. These spots are what I used to measure the possible height of sasquatch seven. This sasquatch also appears to have a more pointed head than the female I saw the evening before. Note that, although its head has a bit of a point at the top of the crown, it does not seem as conical in shape as many ape genera.

Again, the video from which this photo came was shot at 6:45 a.m. I was still sleeping at that time. I arose at 8:00 a.m. that morning, and all the visitors had left the area. Bye, sasquatches.

The sasquatch in charge clearly showed the whites of its eyes while looking at the camera. I believe this one was elderly, and the photo seems to bear this out. Too, the video on which it was captured left me with the impression of an aging sasquatch.

I also believe it carried the young one (sasquatch eight lying in the pines to its right) because in the video, it appeared to bend over, pick up the young one, and hold it facing the camera. My sense is that the elder was trying to show me that the young one had something wrong with it.

Did it want me to help it somehow? I don't know. Much to my chagrin, it wasn't until they were long gone that I had time to peek at the camera pics and videos. This was my last day in camp and I had to break camp, hook-up to the truck and be ready to leave for home early the following morning.

Frankly, the head and eyes of sasquatch seven in photo 13.20 look to me a lot like induced transparency of one type or another. Using the spots on the pine tree just to the right of it, I measured his height at 10 feet 4 inches. His prints measured **28 inches long, $6\frac{7}{8}$ inches at the heel, and $6\frac{1}{4}$ inches at the midpoint**. This massive entity is the largest I'd ever seen! I won't try to guess its weight, although I suspect it was well over 1,100 pounds.

One swipe of its mammoth hand would easily force me to have a bad night! Yet, neither it nor I showed fear or angst of any sort while we were within 52 feet of one another. What in the world was going on?

Stated as simply as possible, when these creatures visit my campsite, I engage in an act of intentional coherence. The act of intentional coherence creates a serene balance within me. This radiates outward from me to the sasquatches. They respond in kind to what I direct outward. My point is simple: Coherence allows positive interactions with sasquatches! More on this appears in the final section of this chapter.

For now, photo 13.21 shows the sasquatch moving to his left toward the female of location 6. The large sasquatch walked over to her, they communicated, and he returned to pick up the young one and hold it toward the camera trap as seen in photo 13.21. Looking carefully at photo 13.21, you should spot the sasquatch's right arm, part of its face, and part of its ankle and foot at the bottom of the photo. Again, partial induced transparency seems clear to me in this photo. Goodness!

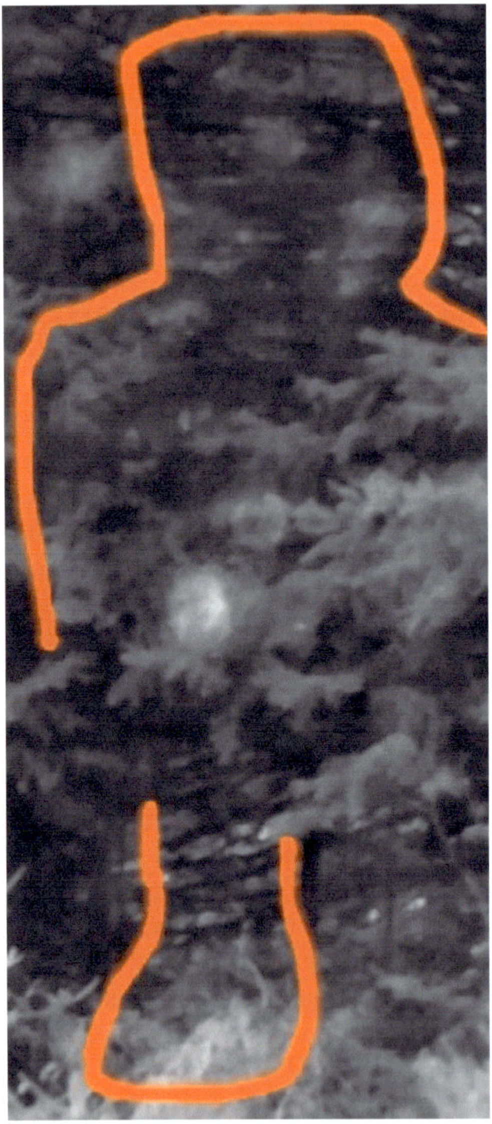

Photo 13.21: Large sasquatch; partial induced transparency?

Photo 13.22 shows the sasquatch bending to pick-up the young one. Audio on the video produced a grunt, like one would make when picking up something heavy.

Photo 13.22: Sasquatch bending over to pick up young one

Photo 13.23 shows the head of the sasquatch from location eight. As mentioned earlier, the large sasquatch bent over, picked this one up, and held it so its face was toward the camera trap in the trailer window. I totally get how challenging it is to see the face of this young one. Even so, patience combined with vigilant observation should help you make out some of its features.

For instance, look for its eyes and mouth; the mouth is long and partially open, with symmetrical teeth visible. Its eyes are wide open, and I perceive it to be experiencing pain or hoping for some type of aid. Sad! Julies' and my hearts were saddened to see this picture! Wish I could have stayed another day or two. Maybe the large sasquatch would have brought it back.

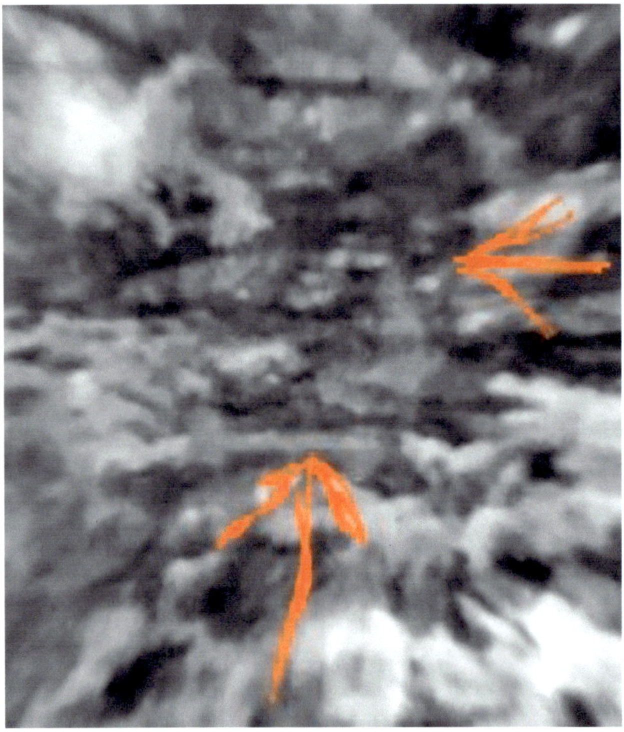

Photo 13.23: Young sasquatch face

The arrow pointing up from the bottom points at the mouth. The arrow pointing in from the left points at an eye.

Photo 13.24 is the final blob photo I use in this book. The marked area encloses the left arm of the largest sasquatch—number seven—as he reached for and took the apple that I left on the small table for him! Recall that the table was 10 feet from the trailer door. The video from which I extracted this photo was obviously filled with circular plasma particles.

Maybe if I sent the video to a professional editor, I might one day have a clearer view of the arm reaching for the apple and the hand taking it. Even so, I can discern the arm from the photo as it appears here only because I've seen the video many times. Looking carefully, you should also be able to see the table at the bottom of the photo just inside the drawn lines. Too, you might detect how the forearm looks smaller than the upper arm. A typical expectation, it seems to me.

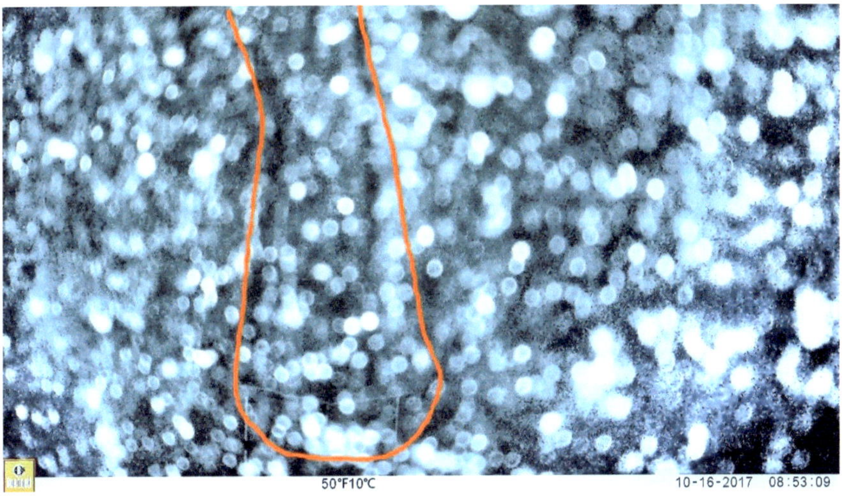

Photo 13.24: Sasquatch reaching for apple on table

Anomalous Experience, Cognitive Dissonance, and Coherence

This seems a proper place to expand on the concepts of *anomalous experience*, *cognitive dissonance*, and *coherence*. Why?

1. Each were present during my extended 2017 stay and, certainly during the experiences of my final night in camp.
2. Cognitive dissonance was so prevalent and in my face near the end of this trip, the psychological factors underlying it absolutely had to influence my thoughts, emotions, and interpretations of my experiences.
3. I believe that practicing coherence accounted—at least, in part—for these amazing experiences, as well as my interpretations.

How Coherence Plays a Crucial Role in Sasquatch Research

I've mentioned something called coherence many times throughout this book. Effective, functional, intentional, and directed coherence plays an *enormous* role in the way Julie and I interact with sasquatches. As such, it's important to understand, at least a little, how coherence works and why we make nearly religious use of it.

I place great emphasis on the act of coherence enhancing my interactions with sasquatches. Remember, coherence as described in this book stabilizes thoughts and emotions while bringing balance to the physical body's internal workings. That is, intentional coherence aids in controlling one's heart rate variability (HRV) pattern. A balanced HRV pattern aids in controlling emotions and thoughts through balancing the autonomic nervous system. Controlling emotions and thoughts also helps one to control and guide one's electromagnetic heart field outward into the surrounding area using directed intent. By the way, the foregoing has been scientifically authenticated many times in diverse situations, conditions, and environments.

Although I lack precise empirical evidence for this next piece, I have phenomenal experiences to confirm the premise. Directing the heart field emanating outward from my body grants a certain amount of control over the way a sasquatch receives it. For example, the heart field is scientifically confirmed; my heart field emanates outward from my body in the form of electromagnetic waves. Sasquatch is going to receive these energy waves regardless of what I do because it is an automatic action. Using directed intent allows me a sense of control. Time for another proposal:

Proposal 9:

> When I engage in intentional coherence, I choose precisely how and where the electromagnetic energy waves emanating from my heart direct outward toward sasquatches.

This is a powerful way to communicate with a creature that seems part human and part animal. Silly as it might sound, I always start my coherence exercise by directing appreciation, compassion, forgiveness, humility, understanding, and valor toward the sasquatches and my surroundings. From there, the situation governs what I do next.

To date, and without fail, sasquatches have expressed a positive response to the electromagnetic energy waves I direct outward from my heart. Yes, they hold strongly to their caution; yes, they stay hidden most of the time; and yes, they have chosen to prevent their young from interacting directly with me, although, youth being youth, I have had a couple of experiences where a young one made sly attempts to get my attention. Darn youngsters, anyway! Clearly, no respect for their elders, eh? Apparently, this occurs across species!

I discuss a bit of the science affecting coherence near the end of Chapter 15, including a brief dialog on a specific aspect of it: HRV. I also detail a coherence exercise Julie and I use to gain and sustain coherence while interacting with sasquatches. Finally, I address the ideas of anomalous experience and cognitive dissonance in relation to coherence in Chapter 15. Next up, though, is a discussion on magnetism, geomagnetism, and sasquatches. The science related with magnetism and geomagnetism, although intense at times, is crucial to my research.

Chapter 14

Geomagnetism — Friend of the Sasquatch?

"Solar activity is the dynamic energy source behind all solar phenomena driving space weather."

— S. K. Pandey and S. C. Dubey

By this time, you are aware of my belief that magnetism and geomagnetism play significant roles in the behavior of and, research for, sasquatches. This chapter begins by relating the basics of magnetism, then describes Earth's core, geomagnetism, Earth's magnetosphere, magnetoreception, magnetite, biomagnetism, magnetic reconnection, and space weather. I lay a foundation for my beliefs by suggesting these individual variables interact as an integral whole to produce potential unusual happenings with sasquatches.

The Basics of Magnetism

The basics of magnetism are not complicated; hence, this is a great starting point. The simplest form of showing magnetism in any context always uses an iron bar with magnetized ends or poles—as seen in diagram 14.1.[70]

Diagram 14.1 shows lines drawn between the top and bottom of the bar magnet as seen in the center of the diagram. The top of the bar magnet is labeled *N* and the bottom *S*. These letters denote the poles of the bar magnet as north and south, or, if you will, they depict different polarities. Circular lines drawn between the N and S designators signify magnetic fields between the north and south poles of the bar magnet. Lines in the magnetic fields vary in strength; the closer to the magnet a magnetic field line is, the stronger the field. Although magnetic fields are invisible to the physical eye, their qualities may be determined using proper instrumentation.

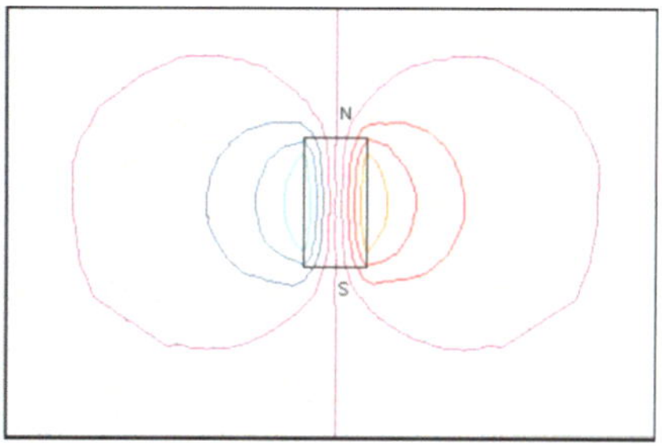

Diagram 14.1: Bar magnet — Credit: NASA

Probably one of the most common attributes associated with magnets coincides with the polarity mentioned above. That is, most folks know that with magnets, like polarities repel, while unlike polarities attract. Next, we'll take a brief peek at Earth's inner core to get a sense for what creates and sustains Earth's magnetism and magnetosphere.

Earth's Inner Core Creates Earth's Magnetosphere

This section discusses how Earth's magnetosphere is formed and supported. Diagram 14.2 depicts a cutout diagram of Earth, showing a solid inner core, liquid outer core, mantle, and crust. Each of these layers play a role in the way Earth's magnetism begins life deep within the bowels of Earth before emanating far out in space to create a protective bubble known as Earth's magnetosphere. Earth's magnetosphere is crucial to life on Earth. In fact, if Earth's core did not generate and sustain Earth's protective magnetosphere, violent energy from the sun would have long since destroyed our beautiful blue planet.

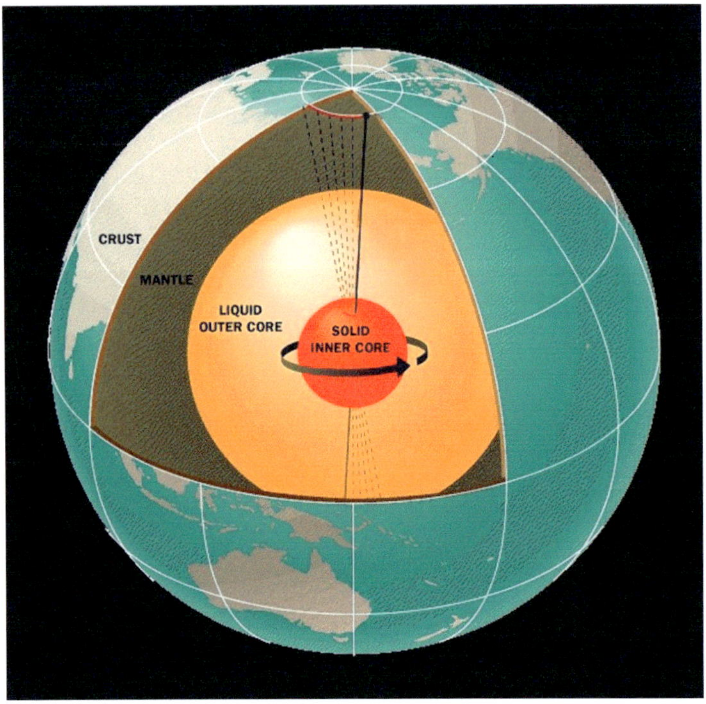

Diagram 14.2: Earth's inner core — Credit: JPL/NASA

Next on deck is a peek at geomagnetism and the way that Earth's magnetosphere protects Earth from dangerous particles from the sun slamming through space and wreaking total havoc on the surface of Earth as well as deep underground. In short, the magnetosphere serves to protect Earth from the sun's damaging particles. Even so, the magnetosphere becomes quite disturbed when extremely fast-moving particles from the sun bombard it. When fast-moving particles from the sun disturb the magnetosphere, Earth and all living organisms on Earth experience its effects. This, then, leads to a discussion on geomagnetism.

Geomagnetism and Earth's Magnetosphere

Odd as it might seem at first blush, in nearly all cases, entire biological systems can be altered by both internal and external magnetic changes, including changes caused by geomagnetism.[71] Reminder: the previous statement is scientific fact and has been proven over many years through repetitive studies conducted by the HeartMath and other private and public institutions. For your information, I rely heavily upon years of research conducted by the HeartMath Institute, although I offer many citations from myriad sources that support their research.

HeartMath's lead researcher, Rollin McCraty informs us that a so-called coupling occurs between human nervous systems and the many types of external, very low geomagnetic frequencies (energy waves) known to propagate during space weather events. In other words, science has consistently confirmed that these types of energy waves interact with the human brain, cardiovascular system, and the autonomic nervous system.[72] In short, space weather events trigger geomagnetic disturbances within Earth's magnetosphere. This, then, is the geomagnetism that relates to my research, and geomagnetism is crucial to many of the proposals I offer throughout the book.

Stated simply, the study of Earth's magnetic field is called the study of geomagnetism.[73] Jeffrey J. Love, Research Geophysicist/USGS Advisor for Geomagnetic Research, put forth:

> Geomagnetism is the study of the earth's magnetic field. This includes the fields produced by the earth as well as those interacting with the earth. Internal dynamo processes within the earth [Earth's inner core] create slowly changing magnetic fields.[74]

Love's statement informs that magnetic fields on Earth's surface can measure different values from location to location and experience changes in values over time. The idea of changing values of Earth's surface magnetism is very significant to my earlier heart-head research as well as to my ongoing sasquatch research, so please do keep it in mind. Love also informed us that some of these changes are slow, such as the centuries-long drift in the direction of the magnetic field, whereas other changes are quick, such as a sudden enhancement during a geomagnetic storm that can occur in minutes.[75] This evidence interweaves directly with my suppositions about geomagnetism and sasquatches.

The National Oceanic and Atmospheric Administration's (NOAA) Space Weather Prediction Center (SWPC) defines Earth's magnetosphere as, "The magnetosphere is the region of space surrounding Earth where the dominant magnetic field is the magnetic field of Earth, rather than the magnetic field of interplanetary space. The magnetosphere is formed by the interaction of the solar wind with Earth's magnetic field."[76] Diagram 14.3 depicts the complexity of Earth's magnetosphere as portrayed by the National Aeronautics and Space Administration (NASA)[77]. Here, I focus on how disturbances to the magnetosphere affect humans, sasquatches, and the Laurentian Shield.

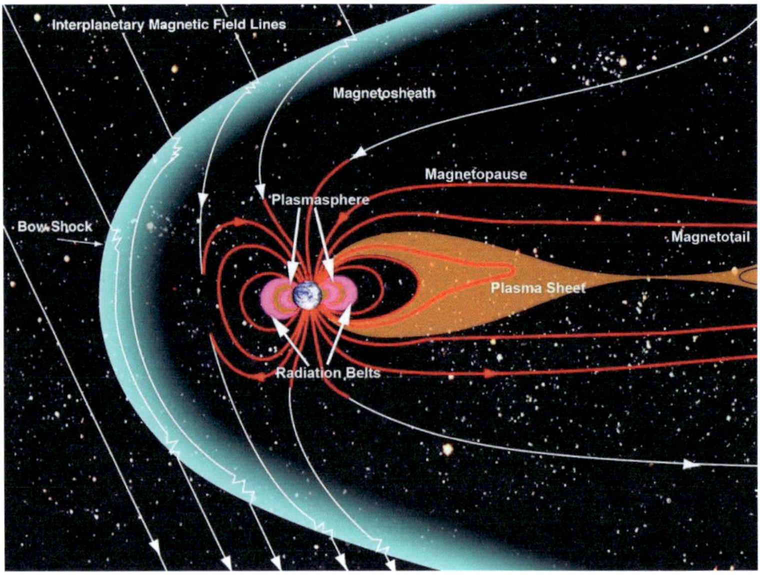

Diagram 14.3: Earth's magnetosphere — Credit: NASA

As a reminder, diagram 14.1 showed a bar magnet complete with a magnetic field. Next, we look at the way Earth's magnetic qualities closely resemble that of a simple bar magnet. This idea is seen in diagram 14.4 as presented by NASA.[78] Note that diagram 14.4 has a dipole with magnetic field lines overlaid on a picture of Earth. NASA did this to enforce the idea that Earth serves as a dipole magnet. Diagram 14.4 might seem like a reproduction of diagram 14.1 and, in a sense, it is. The point here, though, revolves around the idea that Earth *acts* as a magnetic dipole, very similar to a simple bar magnet.

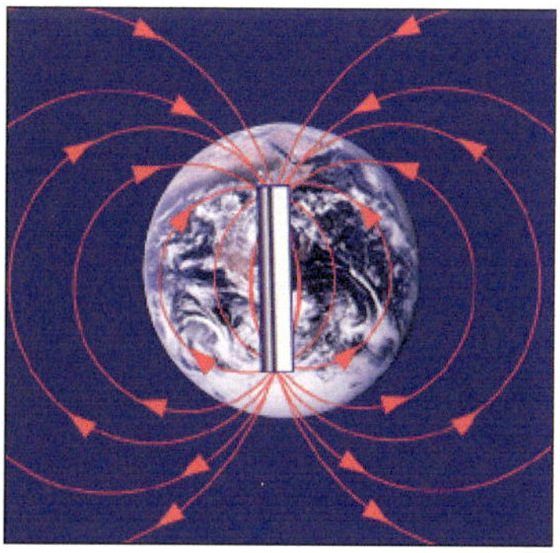

Diagram 14.4: Earth as a bar magnet — Credit: NASA

Now that we know Earth serves as a dipole magnet, the next logical step is to learn a bit more about how our Sun's behavior affects Earth's magnetosphere. After that, we'll look at how solar storms perturbing Earth's magnetosphere touches the health of all living organisms, power grids, underground pipes, drilling, and so forth. Finally, we'll peek at the way geomagnetic storms affect specific geographic areas on Earth—such as the Laurentian Shield—which was introduced in Chapter 2. You may also recall that Chapter 2 discussed the need of dipoles to make the ELF projects work the way they did. Next on deck is a discussion on the interplanetary magnetic field (IMF), which serves as the line of travel for solar storms between the sun and Earth.

The Interplanetary Magnetic Field (IMF)

In a considerably basic sense, the IMF is the sun's magnetic field. This field extends outward from the sun to encompass all the planets within our solar system. Figure 14.5 portrays the IMF[79] as a three-dimensional current sheet; this sheet is now commonly referred to as the heliospheric magnetic field (HMF).

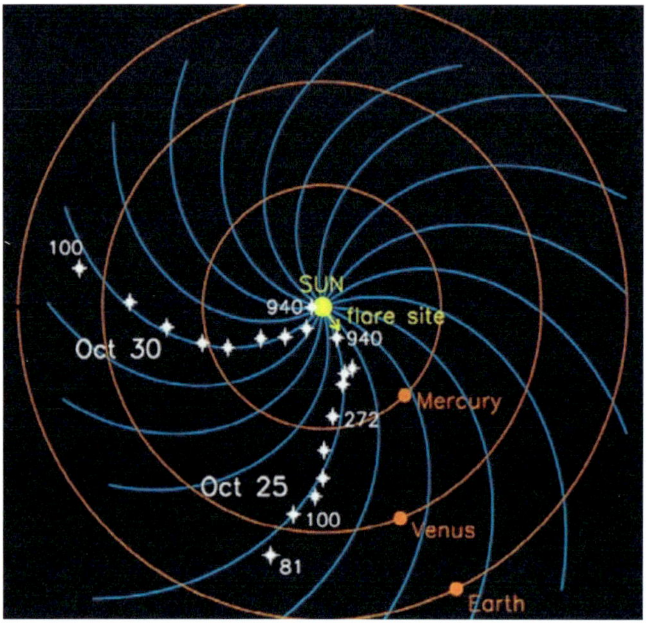

Diagram 14.5: Depiction of the IMF — Credit: NASA

The IMF is the reason solar wind interacts with Earth's magnetosphere.[80] In brief, the IMF is composed of charged particles pulled from the sun's magnetic field. For instance, NASA's STEREO Learning Center notifies us that:

> … structures in the solar wind tend to form a spiral pattern, like water shooting out of a rotating sprinkler head. As with the water drops coming out of the rotating sprinkler, the solar wind particles themselves move straight outwards. However, structures extending across the solar wind flow from a spiral

> pattern extending down to the point on the sun from which they originally came.[81]

To be fair, some scientists disagree with that idea. Even so, NASA also informs us, "Super-fast charged particles called solar energetic particles (SEPs) move much faster than the solar wind and because they are charged, move *along* rather than *across* magnetic field lines. As a result, they trace out the spiral pattern of the IMF" (author's emphasis). [82]

Scientists S. K. Pandey and S. C. Dubey wrote, "Solar activity is the dynamic energy source behind all solar phenomena driving space weather."[83] They also claim that the "… strength of [the] interplanetary magnetic field and its fluctuations have been shown to be [the] most important parameters affecting [Earth's] geomagnetic field variations."[84] These statements advise us that the IMF's role in space weather is critical to understand if we are to grasp the nature of geomagnetic storms on Earth.

I say this mainly because particles of matter and electromagnetic radiation, or billions of tons of plasma, spew off the sun during coronal mass ejections (CMEs), filament eruptions, and solar flares. This plasma then moves along the IMF to eventually interact with Earth's magnetosphere. More on the importance of plasma with respect to sasquatches appears in Chapter 15.

CMEs interact with the solar wind and interplanetary field.[85] They can upset radio transmissions, cause damage to satellites, create electrical transmission line disturbances, and incite health issues. Such disruptions and disturbances result in potentially massive and long-lasting power outages. The important point to remember here is that the IMF serves as the highway between our Sun and Earth for the tons of solar particles blasted off the sun during solar eruptions of any kind. Next, we'll jump right into how space weather affects Earth's magnetosphere, and subsequently, Earth and all living organisms on Earth.

How Space Weather Affects Earth's Magnetosphere

Space weather begins on the sun and travels to Earth and all the other planets within our solar system. Our interest in the sun's role in space weather, however, exists within the framework of its *effects* on Earth and all living organisms that live on Earth.

The sun's heliospheric current sheet plays a primary role in the way space weather travels to Earth. By definition, a current sheet:

> ... is an electric current that is confined to a surface rather than being spread through a volume of space. The heliospheric current sheet results from the influence of the Sun's rotating magnetic field on the plasma in the interplanetary medium. [86]

In addition, NASA tells us that solar storms travel along the sun's current sheet (refer to figure 14.6). NASA also wrote:

> ... small electrical current[s] flow within the [current] sheet. Due to the tilt of the magnetic axis in relation to the axis of rotation of the sun, the heliospheric current sheet flaps like a flag in the wind... As Earth orbits the sun, it dips in and out of the undulating current sheet. On one side the sun's magnetic field points north (toward the sun), on the other side it points south (away from the sun). South-pointing solar magnetic fields tend to cancel Earth's own magnetic field. Solar wind energy can then penetrate the local space around our planet and fuel geomagnetic storms.[87]

You may recall from Chapter 4 that I pointed out the so-called Bz unit of the IMF, and how, "When Bz is southward, or antiparallel to the earth's magnetic field, geomagnetic disturbances become much more severe than when the Bz unit is northward."[88] This information explains quite nicely why we keep track of the Bz unit as we monitor geomagnetic activity. NASA's portrayal of an artist's rendition of the heliospheric current sheet is shown in figure 14.6.

Diagram 14.6: Illustration of Current Sheet — Credit: NASA

The sun is in the middle, and the planets rotate around the sun as seen in figure 14.6. Pluto is the last planet shown in the near front-side of the depiction. Earth, of course, is the third planet from the sun. Next, let's discuss a bit more about solar storms.

The result of solar storms on Earth is measurable and scientifically verifiable.[89, 90, 91] Next, the Natural Environment Research Council [NERC] wrote of space weather, saying:

> During times of heightened space weather, intense solar flares and associated plasma clouds are expelled from the sun. Known as coronal mass ejections (CME), these magnetic clouds can sometimes head directly towards the earth hitting the earth's magnetosphere around 1-3 days later. This will result in a geomagnetic storm…[92]

This is the point where we elaborate on ideas presented earlier in the book. For example, while we expanded our knowledge of the Bz element above, it's now time to reexamine geomagnetic indices in a bit more detail. First, recall that we kept track of the so-called Kp index during field research trips. As a reminder, the Kp index is an indicator of disturbances occurring in Earth's magnetosphere once ejections of particles and plasma from the sun hit this protective bubble.

Keeping track of the Kp index allows us to record such disturbances, thereby setting a stage for documenting sasquatch behavior before, during, and right after a geomagnetic storm. This idea is pertinent because our observations strongly point to how sasquatch behaviors alter from baseline behaviors—where baseline behaviors are defined as behaviors happening when no disturbances to the geomagnetic field occur. As I've stated before, I have several years' worth of observational evidence supporting this premise.

NOAA's SWPC monitors the Kp index daily, with updates occurring every three hours. For example, scientists Pandey and Dubey wrote, "… values of Kp index are available on a 3-hour interval logarithmic scale."[93] This, in turn, allows for a 3-hour precision factor as we observe changing behaviors with sasquatches. This precision factor notion is important to our overall theme that space weather affects sasquatches; I delve into it in more detail in the next section.

Before getting to that, though, I wish to discuss how NOAA's SWPC shares information on a given space weather incident. Good fortune! We can examine a real-life geomagnetic storm as it is occurring!

Right now, on the date I'm writing this paragraph, April 20, 2018, there is a minor geomagnetic storm occurring. Figure 14.7 shows the SWPC's warning graphic for this storm.[94]

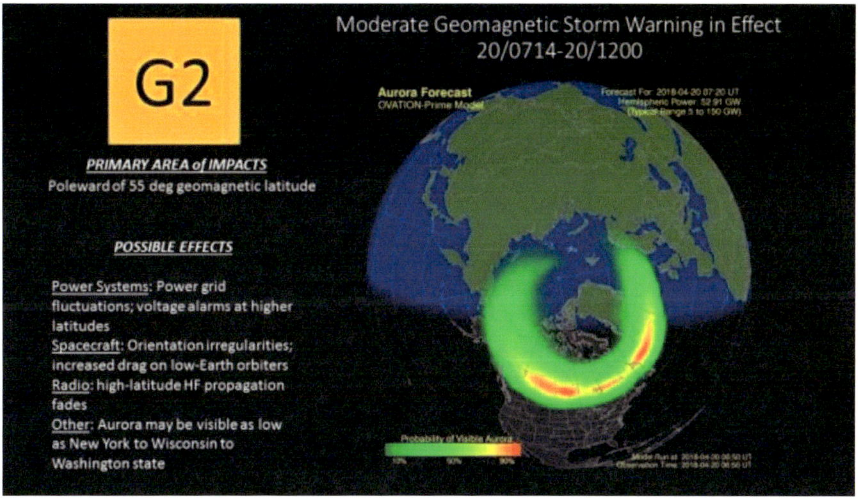

Diagram 14.7: Storm warning graphic — Credit: NOAA

Diagram 14.7 illustrates specific data about the current geomagnetic storm, including possible effects of the storm on Earth. These effects include:

1. **Power systems:** High-latitude power systems may experience voltage alarms, and long-duration storms may cause transformer damage.
2. **Spacecraft operations:** Corrective actions to orientation may be needed by ground control and possible changes in drag affect orbit predictions.
3. **Radio:** High-latitude HF [high frequency] radio propagation can fade.
4. **Other systems:** Auroras may be visible as low as New York to Wisconsin to Washington state.[95]

From this information, it is easy to see that today's (April 20, 2018) G2 geomagnetic storm produces effects not only throughout our atmosphere, rather, also on the surface of Earth! A question related to the knowledge that this day's geomagnetic storm is affecting things on Earth might be, "How do I get from here to the idea that geomagnetic storms produce behavior changes in sasquatches?" Hmm! We are getting closer and closer to realizing an answer for that question. Please hang with me for just a bit longer.

Geomagnetic Storms and the Laurentian Shield

Geomagnetic storms also affect the functional nature of the Laurentian Shield (introduced in Chapter 2). Recall that the Laurentian Shield is an "underlying formation of old, dry granite"[96] found under northern Wisconsin and the western Upper Peninsula of Michigan. It covers a very large geographical area and is made up of low conductivity rock, which means electromagnetic waves easily flow through it. Further, granite is considered one of many natural sources for non-ionizing radiation. I discuss the ideas of ionizing and non-ionizing radiation in Chapter 15.

My interest in the Laurentian Shield relates to how extremely low frequency waves flowing through the ground affect experiences with sasquatches. We learned in Chapter 2 that the U.S. Navy made practical use of the Laurentian Shield by setting up two ELF facilities; one in Republic, Mich., and the other in Clam Lake, Wisc. The transmitters in each location allowed them "to work in concert, as [one] giant antenna, with [two] probes 148 miles apart, still a mere fraction of the generated wavelength."[97] The Laurentian Shield rock formation is relatively close to the surface and, as mentioned above, covers a large physical area.

As a note of interest, few places in the United States meet the conditions of being close to the surface with low conductivity. In fact, Carlos Altgelt, a Buenos Aires, Argentina broadcast specialist, claims Wisc. and Mich. are the best two locations within the United States that satisfy these two conditions.[98] This, of course, is the primary reason the U.S. Navy selected these areas for its ELF project so many years ago. This is also why these areas are *naturally* conducive to electromagnetic waves (light waves) flowing through them, which, in turn, affects sasquatch behaviors.

Because the Laurentian Shield serves as a conduit for low frequency waves, geomagnetic activities within its boundaries are readily measured. Furthermore, whenever Earth's magnetosphere is perturbed—whether slightly or greatly—electromagnetic (light) waves flow through it. Based on this information, I offer two more proposals:

Proposal 10:

> The geology of the Laurentian Shield offers sasquatches a natural geomagnetic/electromagnetic habitat in which to wander and engage in typical sasquatch behaviors.

Proposal 11:

> The geology of the Laurentian Shield offers a way for sasquatches to interact with and manipulate extremely low frequencies to their benefit.

Sasquatch researcher Pearl Prihoda's Photosynthetic Piezoelectric Induced Transparency (PPIT) theory addresses geographic locations where natural Earth magnetism changes of the ELF variety are readily available. For example, she believes such changes occur near seismic fault lines and areas where the bedrock geology introduces natural Earth magnetism changes. The Laurentian Shield is clearly one such geographic zone, and it covers an enormous amount of land area between the western Upper Peninsula of Michigan and the Northern Highland geographical area covering most of northern Wisconsin. This includes the 1.5 million plus acres of the Chequamegon-Nicolet National Forest!

As a reminder, studies show definitive alterations in carbon-based humans during "sharp or sudden variations in geomagnetic and solar activity..."[99] Furthermore, Chairman of the Board of the Saudi Telecom Company Abdullah Alabdulgader, along with HeartMath's lead researcher, Rollin McCraty and five other scientists, put forth evidence in a February 2018 *Scientific Reports* article claiming that geomagnetic storms "alter regulatory processes such as melatonin/serotonin balance, blood pressure, breathing, reproductive, immune, neurological, and cardiac system processes."[100] Data from this report is quite significant to my sasquatch research and, although it is a bit geeky to read, it is well-worth a look-see.

Moreover, many scientists claim geomagnetic disturbances associate with "significant increases in hospital admissions for depression, mental disorders, psychiatric admission, suicide attempts, homicides and traffic accidents." [101] My goodness! Also, it is commonly known that when Earth's magnetosphere is disturbed during space weather events, existing diseases can be aggravated, along with significant increases occurring in cardiac arrhythmia, cardiovascular disease, incidence of myocardial infarction-related death, alterations in blood flow, increased blood pressure, and epileptic seizures.[102] In other words, space weather directly affects living organisms and not always in positive ways. This information is also important to keep in mind as we move forward because I believe that sasquatches—as carbon-based creatures—are also affected during geomagnetic disturbances!

Next up, I address magnetoreception, a fascinating and scientifically validated event affecting humans, animals, and essentially all living organisms on Earth.

Magnetoreception

Wikipedia defines magnetoreception as a "... sense which allows an organism to detect a magnetic field to perceive direction, altitude or location."[103] Magnetoreception, although still somewhat controversial among some scientists, is an extraordinarily interesting topic and, one which might lead to a better understanding of certain unusual happenings occurring with sasquatches. For example, a 2011 article published online at *LiveScience*, reports, "Humans may have a sixth sense after all, suggests a new study finding that a protein in the human retina, when placed into fruit flies, has the ability to detect magnetic fields."[104] If so, sasquatches, commonly thought to be hairy, human-like apes, might be included within the ranks of magnetoreceptive mammals.

Roswitha Wiltschko's research group at Goethe University Frankfurt, along with French colleagues, figured out that birds have a light-dependent compass in their eyes. Their research results showed, "This [light-dependent] compass [in their eyes] gives them information about the direction of the earth's magnetic field ."[105] In addition, the study also revealed:

> Birds have sensory organs for orientation and navigation in the earth's magnetic field: They use their beak to measure the strength of the magnetic field, while their eyes provide directional information. One type of cone photoreceptors in the birds' eyes is sensitive to UV light and also contains a form of the protein cryptochrome. Previous studies of the Frankfurt researchers suggested that most likely it is this that enables birds to detect the magnetic field.[106]

Cryptochrome is considered a prominent factor in magnetoreception, possibly even relating magnetoreception to humans. More on this in just a moment.

A 2013 article published online at *The Scientist: Exploring Life, Inspiring Innovation*, reported, "Researchers from various disciplines are homing in on the mechanics of magnetoreception, an enigmatic sense that some animals use to navigate the globe."[107] The same article, titled, "A Sense of Mystery," similarly reported, "… scientists have found that animals also use the earth's field lines as *magnetic 'signposts'—positional information* created by unique combinations of field inclination and intensity at *specific geographic locations* (author's emphasis)."[108]

This idea is certainly thought-provoking to say the least! And well worthy of reflection when evaluating the possibility that sasquatches might intentionally interact with Earth's magnetism. Think about this for a moment! Can you imagine Earth's magnetic field lines serving as **signposts** to sasquatches? Maybe this accounts for some of the stick sign pointers sasquatch researchers find in the wilderness. Or, perhaps Earth's magnetic energies served as a signpost when four or more sasquatches turned a beautiful young maple tree into an arched directional pointer as mentioned in Chapter 5. Are you still open-minded? Great! Then let's continue.

The "A Sense of Mystery" article continued, suggesting that, "The existence of the signpost sense and the internal compass are now generally accepted, and a number of scientists think that some animals may have both."[109] Startling, wouldn't you say? Or, perhaps, spectacularly revealing to say more.

Furthermore, the same article proposed:

> Most researchers in the field agree that the compass sense is likely seated in cryptochromes within the eye, and many are convinced that there is another sense, most likely a signpost sense, passed through the trigeminal nerve [a nerve responsible for sensation in the face and motor functions such as biting and chewing[110]] and probably based on some sort of iron-containing, magnetism-sensing cells in beaks or snouts.[111]

Interesting, isn't it, how we once again bump into the idea of magnetism. This offers added and, perhaps, profound support to the idea that sasquatches may be able to *literally* feel (sense) changing Earth magnetism, thereby making intentional and controlled use of it. Particularly, perchance, during those times when Earth's magnetosphere is disturbed from solar storms.

In yet another online study, "The Radical Search for a Magnetic 'Sixth-Sense' in Humans" posted online July 3, 2016, Shelly Fan, a neuroscientist at the University of California, San Francisco who studies ways to make old brains young again, put forth:

> [A] magnetic sense doesn't just serve migratory animals. Among biologists, it's now widely accepted that *many species* are attuned to the earth's magnetic field. Among these are surprising candidates such as

> bacteria, lobsters, and *mammals* including wood mice and dogs (author's emphasis).[112]

The report continued, suggesting:

> *Cryptochromes*, which sit in the eyes, likely measure changes in the [magnetic] field's *intensity* and *inclination* — that is, how sharply the magnetic field's angle strikes the earth's surface — rather than the *state* of the field directly. The even more mysterious magnetite [Fe_3O_4] may be present in other organs related to navigation, such as the inner ear. *Both systems could work together* to provide a spherical coordinate system used for spatial orientation (author's emphasis).[113]

Hmm … perhaps use of the protein cryptochrome is an innate action, hence, what I perceive as intentional could very well be a natural act? The puzzle continues. Conceivably, this might mean that sasquatches make use of Earth's magnetic fields without cognitive awareness. I really do not know at this point. Yet another reason research continues, eh?

Furthermore, scientists offer a relevant reason why magnetoreception is difficult to study in humans. They suggest that if magnetoreception exists in humans, "it's likely processed subconsciously."[114] Clearly, studying the subconscious is a daunting task. If, however, a sense of magnetoreception does occur in humans and is then processed within the subconscious, I suspect something similar might also be true with sasquatches. Which, I suppose, could make interacting with Earth's magnetic field a natural act. Startling to consider, yet worthy of thought?

There seems to be a clear intent associated with each so-called sasquatch stick marker we've come across over the years. Perhaps an innate directional sense drives placement of markers and pointers in certain geographic locations, and many of these markers have been found to follow correct compass directions. Who knows for sure? I might add to the above by pointing out that many woods professionals over the years have found their way in wildernesses around the world without use of a compass. I have seen this done. Maybe folks with this innate ability key into a natural magnetoreceptive skill, process it subconsciously, and then wander the wilderness without fear of ever getting lost. For sure and certain, there is a lot about the quality of magnetoreception we do not yet know.

Another example of natural sensing relates to martial arts. Many martial artists train all their senses to detect threats (something coming at them) or other changes in their environment. Following years of training, they engage in this activity successfully—often while blindfolded. Can we prove they are **not** using an innate magnetic field identifier skill? Hmm! Remember, movement within a magnetic field causes changes in the field. Changes in these fields produce sounds. It might indeed be changes in Earth's magnetic fields causing sounds that such skilled martial artists detect. If so, would it be a huge leap in logic to associate a similar skill to sasquatches? I say no, not if they are carbon-based creatures like humans.

Continuing, Robert Gegear, head of the Gegear Laboratory at Worcester Polytechnic Institute, co-authored an intriguing 2008 study, "Cryptochrome Mediates Light-dependent Magnetosensitivity in *Drosophila* [a genus of flies, belonging to the family *Drosophilidae*, whose members are often called small fruit flies or pomace flies, vinegar flies, or wine flies[115]]." The study concluded that:

> The ability of an animal to detect geomagnetic fields has substantial biological relevance as it is used by many invertebrate and vertebrate species *for orientation and navigation purposes, including homing, building activity and long-distance migration* (author's emphasis).[116]

Finally, this compendium of information leads me to another proposal:

Proposal 12:

> The way magnetic senses have been proven to aid mammals and insects may be applied fairly and confidently to sasquatches.

Although I cannot offer empirical evidence to substantiate all my thoughts on this topic, I believe quite strongly that the theme of magnetoreception and sasquatches is open for serious discussion and ongoing research. I suspect that geomagnetic storms present sasquatches with a greater ability to manipulate Earth's magnetism to their benefit than when Earth's magnetosphere is totally quiet. More on this idea a bit later.

Furthermore, I have in mind the idea that sasquatches manipulate Earth's magnetism to interfere with electronic devices. For instance, this might account for so-called blob photos presented by many members of the sasquatch community, especially those including what appear as plasma blotches or circles within so many blob photos. This could also underlie witness stories about folks having electronic issues with vehicles during sasquatch encounters. Simply another thought to consider as part of a whole.

I know I've spent a bit of time on the premise of magnetoreception; however, I believe this concept is important enough to our story that the amount of time spent with it speaks for itself. Prior to moving into a discussion on biomagnetism, I address the mineral magnetite—a magnetic nano-sized crystal found in humans and animals.

Magnetite

Catriona Houston, a writer who holds a PhD in the modulation of inhibitory synaptic transmission, published an article in March 2018 on the blog, "Knowing Neurons," an informative website created by young neuroscientists to creatively share neuroscience information. She wrote, "Controlling the minds of others from a distance has long been a favourite science fiction theme — but recent advances in genetics and neuroscience suggest that we might soon have that power for real."[117] Excuse me?

Houston also wrote about a relatively new technique that uses "low-frequency radio waves or a magnetic field"[118] to carry out a form of mind-control. Hmm! She concluded her article suggesting that:

> … these new systems are potentially more precise and less invasive than existing techniques for altering brain activity such as deep brain stimulation. With so much progress on a variety of fronts, some form of human mind control — and the treatments and benefits it confers — should be here before long. We just need to make sure that like other emerging technologies — artificial intelligence and robotics come to mind — they are used for good to improve lives.[119]

I assuredly support the latter part of this quote! Now let's peek at how low-frequency waves or a magnetic field can alter brain function.

Stated simply, low-frequency radio waves can alter brain function due to nanoparticles in the brain; these particles are nanocrystals known as magnetite. Magnetite is an "iron oxide—Fe3 O4—found in cells of some organisms, which may play a part in geomagnetic orientation because of its magnetic properties."[120] Recent articles suggest that there are millions of magnetite nanocrystals in the human brain. Some say at least five million! While that sounds like a lot, considering the extremely small size of these magnetite crystals, they take up very little space within the brain.

Joseph L. Kirschvink et al., researchers at the Division of Geological and Planetary Sciences, The California Institute of Technology, Pasadena, California discovered in 1992 that magnetite nanocrystals within the human brain can be intentionally affected by introducing weak ELF radio-frequency signals into the nanocrystals.[121] This is certainly not our first mention of ELF, for I have addressed ELF numerous times throughout the book, so this is merely another glance at how the use of ELF radio-frequencies can affect the human body—in this instance, the human brain. Go ahead … let your imagination run wild!

Considering that magnetite is easily magnetized, I put forth the possibility that sasquatches may be influencing humans via magnetic manipulation of the magnetite nanoparticles found within the human brain. Secondarily, I'll point out that the first mention of magnetite came earlier in this chapter under the discussion about magnetoreception and magnetic navigation.

Reto Gieré, a scientist at the Department of Earth and Environmental Science, University of Pennsylvania, Philadelphia, PA, wrote, "The presence of magnetite in humans, however, also has other potential implications, including possible biological disorders linked to the weak magnetic fields generated by cellular phones, electric power lines, and appliances, or high-field saturation effects from exposure to strong magnetic fields during MRI procedures."[122] Perhaps as researchers continue to explore the issue of magnetite in the human brain, additional insights will eventually sprinkle into common awareness.

Regardless, magnetite nanoparticles (crystals) within the human brain have been shown to be easily magnetized with extremely low radio frequencies. To that end, I believe we have proved—and continue to prove—the importance of ELF to sasquatch research.

Deeply delving into the mineral magnetite and the way it is affected by external magnetic fields would involve probing into the way that quantum electron spin affects it. Seriously, probing into a swamp of silliness is way outside the scope of this book. Not only that, although I can drill into that arena to a certain extent, I am not qualified to do it justice. Limitations suck sometimes; even so, I believe it is wise to remain aware of one's limitations. For now, let's say that this brings me to another proposal:

Proposal 13:

> Magnetite nanoparticles within the human brain may be manipulated by sasquatches controlling and directing extremely low frequencies.

Next up—a brief discussion on biomagnetism, which also plays a role in my suppositions about sasquatches, magnetism, geomagnetism, and space weather.

Biomagnetism

Biomagnetism—as used in this book—is defined simply as "the magnetic field created by a living organism, the effect of an external magnetic field on a living organism, [and] the scientific study of these phenomena."[123] Frankly, this definition suffices to explain why I incorporate it here.

I offer evidence later in the book about the way humans and other living organisms generate a measurable magnetic field. I also present data and studies on how an external magnetic field affects living organisms. Finally, I purport that this includes sasquatches. In fact, I believe this is reasonably vital information with respect to unraveling at least one of the mysteries surrounding the sasquatch phenomenon. For example, I offer the following proposal:

Proposal 14:

> As carbon-based creatures, sasquatches emanate a heart magnetic field—as do humans—and, quite possibly, their heart's magnetic field extends farther out than the 10 feet measured on humans.

I lack empirical evidence to support this proposal. Even so, my experiences with sasquatches lead me to believe this is a valid train of thought for continued research.

The natural magnetic qualities of the human biological system interacting with very low or extremely low external frequencies occurring on Earth during space weather events falls under the umbrella of biomagnetism. Understanding how biomagnetism works and interacts with geomagnetism grants a realistic foundation upon which we can build solid evidence revealing how sasquatches might make directed and intentional use of geomagnetism for their benefit. This, then, is an area of research worthy of continued attention as I press ever forward, probing more and more deeply into the sasquatch phenomena.

The final piece I wish to discuss in this chapter deals with something known as magnetic reconnection. Even though this space weather factor probably plays the least role in my assumptions about magnetism, geomagnetism, and sasquatches, I feel it deserves a bit of attention.

Magnetic Reconnection

Believe it or not, a popular science fiction idea has exploded into our present-day reality as a vibrant and real event. It's an idea that has tantalized movie goers and readers of science fiction for many years. It's an awareness so radical, that scientists have always stifled grins, persistent headshaking, and outright laughter each time it was brought up with any verve of seriousness. Not so any more.

There is now ongoing dedicated funding for, and investigation into, a silly science fiction idea known as portals! Yup ... portals. Get outta here, you say? Well, if you learned how to use the relatively newfound portals, you might indeed be able to "get outta here" in ways never imagined! But where would you go? Stated simply, to the sun. Ouch. That might be a bit warmer than you want; no, it would be too hot to stand! Why? Because the sun's surface temperature is commonly thought to be 5,500 degrees C (10,000 degrees F), while farther away from the sun's surface, the temperature climbs to an astounding approximate 2,000,000 degrees C (3,600,000 degrees F).[124] Yikes! Better not book a trip through any of these portals at this point! Seriously!

NASA funded researcher and plasma physicist Jack Scudder of the University of Iowa, wrote, "[Portals are] places where the magnetic field of Earth connects to the magnetic field of the sun, creating an uninterrupted path leading from our own planet to the sun's atmosphere 93 million miles away."[125] NASA also reported that "observations by NASA's THEMIS spacecraft and Europe's Cluster suggest that these magnetic portals open and close dozens of times each day."[126]

In short, these portals result from a phenomenon known as magnetic reconnection. Before you question why I include this information, know this: The study of this phenomenon is so important that NASA launched the Magnetospheric Multiscale (MMS) mission on March 12, 2015. NASA states, "MMS consists of four identical spacecraft that orbit around Earth through the dynamic magnetic system surrounding our planet to study a little-understood phenomenon called magnetic reconnection."[127]

Further, the NASA mission statement for the MMS informs us:

> The Magnetospheric Multiscale, or MMS, mission studies the mystery of how magnetic fields around Earth connect and disconnect, explosively releasing energy via a process known as magnetic reconnection.[128]

NASA also states that, "magnetic reconnection is a phenomenon unique to plasma," and it occurs "when magnetic field lines cross and release a gigantic burst of energy."[129] Similarly, according to Princeton Plasma Physics Laboratory (PPPL):

> [Magnetic] reconnection occurs when the magnetic field lines in plasma—the collection of atoms and charged electrons and atomic nuclei, or ions, that make up 99 percent of the visible universe—converge and forcefully snap apart. Electrons that exert a varying degree of pressure form an important part of this process as reconnection takes place.[130]

Continuing, PPPL's Raphael Rosen wrote, "Physicist Fatima Ebrahimi … has published a paper showing that magnetic reconnection—the process in which magnetic field lines snap together and release energy—*can be triggered by motion in nearby magnetic fields* (author's emphasis)."[131] I believe it is well worth the effort to explore NASA's web material on the MMS mission even though a bit of scientific controversy surrounds the entire magnetic reconnection field. The MMS online home address is: https://mms.gsfc.nasa.gov/index.html.

The point of sharing a bit of information on magnetic reconnection is simple. It offers yet another piece of material about why I believe the sun's interaction with Earth's magnetism is an overpoweringly important factor when researching and studying the habits of sasquatches and their behavior patterns. Please bear in mind that magnetic reconnection—as described above—includes magnetic field lines, changes in magnetic fields, and plasma. These qualities and more are either considered or discussed at some length throughout this book, making inclusion of this information pertinent overall.

It is also important to note that a bit of controversy surrounds the information put forth by NASA and the MMS team. For example, not all scientists believe that this information is as valid as touted by NASA and the MMS team. Regardless, it is, at least to me, fascinating information and while future studies will ultimately prove who is right vs. wrong, we cannot deny that something is happening to cause explosive reactions "when magnetic field lines cross and release a gigantic burst of energy."[132]

Chapter 15 offers a brief excursion into some of the science important to a rational development of my proposals. I recognize that some of the science included may be a tough pill to swallow; however, without it, my beliefs border on pure speculation. My goal here is to share ways science supports my premises on sasquatches, magnetism, geomagnetism, magnetoreception, magnetite, biomagnetism, magnetic reconnection, and possible intentional electromagnetic interactions by sasquatches. Yes, I continue to go bold! First, though, take a brief gander at the closing section in this chapter.

Driving to the Sun?

As an aside, on February 6, 2018, SpaceX CEO Elon Musk launched SpaceX's first heavy rocket. On it, he placed his cherry red Tesla roadster—and it's headed to Mars. The driver of that vehicle is a silly mannequin dressed in a space suit and named "Starman."

We should probably leave the above-mentioned *portal* trip to Elon Musk's cherry red Tesla roadster. That is, unless you believe you are *burn free*! Below is the final photo of Musk's roadster that will ever be seen—unless, of course, an alert alien spots the roadster, snaps unexpected shots, and shoots the photos to Earth.

Photo 14.1: Roadster headed to Mars — Credit: Space.com

Chapter 14

[70] Staff, NASA. "An Educator's Guide with Activities in Physical Science." Mapping Magnetic Influence. Accessed April 12, 2018. https://www.sunearthday.nasa.gov/swac/materials/Mapping_Magnetic_Influence.pdf.

[71] McCraty, Rollin. "Global Coherence Research: Human-Earth Interconnectivity." In *Science of the Heart: Exploring the Role of Heart in Human Performance*, 89. Vol. 2. Boulder Creek: HeartMath Institute, 2015.

[72] Ibid., 89.

[73] Staff. "Earth's Dipole Magnetic Field." UCAR Center for Science Education. Accessed April 12, 2018. https://scied.ucar.edu/earths-dipole-magnetic-field.

[74] Love, Jeffrey J. "Introduction to Geomagnetism." U.S. Geological Survey. Accessed April 21, 2018. https://geomag.usgs.gov/learn/introtogeomag.php.

[75] Ibid.

[76] Staff, NOAA. "Earth's Magnetosphere." NOAA / NWS Space Weather Prediction Center. Accessed April 12, 2018. https://www.swpc.noaa.gov/phenomena/earths-magnetosphere.

[77] Zell, Holly. "Earth's Magnetosphere and Plasmasheet." NASA. March 02, 2015. Accessed April 12, 2018. https://www.nasa.gov/mission_pages/sunearth/science/magnetosphere2.html.

[78] THEMIS. "Earth's Magnetosphere." THEMIS - Time History of Events and Macroscale Interactions During Substorms. Accessed May 17, 2018. http://cse.ssl.berkeley.edu/artemis/mission-mag.html.

[79] Staff, NASA. "Interplanetary Magnetic Field." NASA - STEREO Learning Center. Accessed April 13, 2018. https://stereo.gsfc.nasa.gov/classroom/definitions.shtml#IMF.

[80] SpaceWeatherLive, Staff. "The Interplanetary Magnetic Field (IMF) | Help." SpaceWeatherLive.com. Accessed April 13, 2018. https://www.spaceweatherlive.com/en/help/the-interplanetary-magnetic-field-imf.

[81] NASA. "Interplanetary Magnetic Field." NASA - STEREO Learning Center. Accessed April 20, 2018. https://stereo.gsfc.nasa.gov/classroom/definitions.shtml.

[82] Ibid.

[83] Pandey, S. K., and S. C. Dubey. "Impact of Solar and Interplanetary Disturbances on Space Weather." *International Research Journal of Advanced Engineering and Science* 2, no. 1 (2017): 125-30. Accessed April 19, 2018, 125.

[84] Ibid., 126.

[85] "Coronal Mass Ejection." Wikipedia. April 16, 2018. Accessed April 19, 2018. https://en.wikipedia.org/wiki/Coronal_mass_ejection.

[86] "Current Sheet." Wikipedia. Wikimedia Foundation, September 30, 2018. https://en.wikipedia.org/wiki/Current_sheet.

[87] Zell, Holly. "The Heliospheric Current Sheet." NASA. March 02, 2015. Accessed May 18, 2018. https://www.nasa.gov/content/goddard/heliospheric-current-sheet.

[88] "Space Weather Glossary." Space Weather Prediction Center - NOAA. Accessed March 1, 2018. https://www.swpc.noaa.gov/content/space-weather-glossary#b.

[89] Vencloviene, Jone, et al. "Short-Term Changes in Weather and Space Weather Conditions and Emergency Ambulance Calls for Elevated Arterial Blood Pressure." *MDPI*, Multidisciplinary Digital Publishing Institute, 20 Mar. 2018, www.mdpi.com/2073-4433/9/3/114.

[90] Babayev, Elchin S., and Aysel A. Allahverdiyeva. "Effects of Geomagnetic Activity Variations on the Physiological and Psychological State of Functionally Healthy Humans: Some Results of Azerbaijani Studies." *Advances in Space Research*, Pergamon, 7 Sept. 2007, www.sciencedirect.com/science/article/pii/S0273117707009404?via%3Dihub.

[91] Rozhkov, V. P., M. I. Trifonov, S. S. Bekshaev, N. K. Belisheva, S. V. Pryanichnikov, and S. I. Soroko. "Assessment of the Effects of Geomagnetic and Solar Activity on Bioelectrical Processes in the Human Brain Using a Structural Function." SpringerLink. March 03, 2018. Accessed April 28, 2018. https://link.springer.com/article/10.1007%2Fs11055-018-0564.x.

[92] NERC. "Ground Effects of Space Weather." Ground Effects of Space Weather. Accessed April 07, 2018. http://www.geomag.bgs.ac.uk/education/gic.html.

[93] Pandey, S. K., and S. C. Dubey. "Impact of Solar and Interplanetary Disturbances on Space Weather." *International Research Journal of Advanced Engineering and Science* 2, no. 1 (2017): 125-30. Accessed April 19, 2018, 125.

[94] SWPC, NOAA. "Homepage | NOAA / NWS Space Weather Prediction Center." NOAA Space Weather Prediction Center. April 20, 2018. Accessed April 20, 2018. https://www.swpc.noaa.gov/.

[95] Ibid.

[96] Biologic Effects of Electric and Magnetic Fields Associated with Proposed Project Seafarer. Ft. Belvoir: Defense Technical Information Center, 1977. doi:10.17226/20341.

Note: This report was prepared in 1977 by the Committee on Biosphere Effects of Extremely-Low-Frequency Radiation in response to a request from the United States Navy.

[97] Villeneuve, Bradley. "ELF Station Republic, MI." Military History of the Upper Great Lakes. October 10, 2015. Accessed June 25, 2018. http://ss.sites.mtu.edu/mhugl/2015/10/10/elf-sta-republic-mi/, 5.

[98] Altgelt, Carlos A. "The World's Largest 'Radio' Station." PDF. Accessed 17 February 2018. http://www.hep.wisc.edu/~prepost/ELF.pdf, page 6.

[99] Alabdulgader, Abdullah, Rollin McCraty, Michael Atkinson, York Dobyns, Alfonsas Vainoras, Minvydas Ragulskis, and Viktor Stolc. "Long-Term Study of Heart Rate Variability Responses to Changes in the Solar and Geomagnetic Environment. "*Scientific Reports* 8, no. 1 (2018). Accessed June 21, 2018. doi:10.1038/s41598-018-20932.x, 1.

[100] Ibid., 1-2.

[101] Ibid., 1-2.

[102] Ibid., 1-2.

[103] "Magnetoreception." Wikipedia. April 13, 2018. Accessed April 14, 2018. https://en.wikipedia.org/wiki/Magnetoreception.

[104] Bryner, Jeanna. "Humans May Have 'Magnetic' Sixth Sense." LiveScience. June 21, 2011. Accessed April 13, 2018. https://www.livescience.com/14694.humans-sixth-sense-magnetic-fields.html.

[105] Goethe University Frankfurt. "New insight into the light-dependent magnetic compass of birds." ScienceDaily. Accessed April 16, 2018. www.sciencedaily.com/releases/2016/06/160606100519.htm.

[106] Ibid.

[107] Cossins, Dan. "A Sense of Mystery." TheScientist. August 1, 2013. Accessed April 13, 2018. https://www.the-scientist.com/?articles.view/articleNo/36722/title/A-Sense-of-Mystery/.

[108] Ibid.

[109] Ibid.

[110] "Trigeminal Nerve." Wikipedia. April 10, 2018. Accessed April 20, 2018. https://en.wikipedia.org/wiki/Trigeminal_nerve.

[111] Cossins, Dan. "A Sense of Mystery." TheScientist. August 1, 2013. Accessed April 13, 2018. https://www.the-scientist.com/?articles.view/articleNo/36722/title/A-Sense-of-Mystery/.

[112] Fan, Shelly Xuelai. "The Radical Search for a Magnetic 'Sixth Sense' in Humans." Singularity Hub. March 31, 2017. Accessed April 13, 2018. https://singularityhub.com/2016/07/03/the-radical-search-for-a-magnetic-sixth-sense-in-humans/#sm.0000g0gpqu9hpeyqvpx14fowj4r9r.

[113] Ibid.

[114] Ibid.

[115] "Drosophila." Wikipedia. April 15, 2018. Accessed April 15, 2018. https://en.wikipedia.org/wiki/Drosophila.

[116] Gegear, Robert J., Amy Casselman, Scott Waddell, and Steven M. Reppert. "Cryptochrome Mediates Light-dependent Magnetosensitivity in Drosophila." Nature News. July 20, 2008. Accessed April 14, 2018. doi:10.1038/nature07183.

[117] Houston, Catriona, Ryan Jones, Tessa Abagis, and Sean Noah. "Remote Control of the Brain Is Coming: How Will We Use It?" Knowing Neurons. March 21, 2018. Accessed February 22, 2019. https://knowingneurons.com/2018/03/21/remote-control-brain/.

[118] Ibid.

[119] Ibid.

[120] "Magnetite." The Free Dictionary. Accessed February 22, 2019. https://medical-dictionary.thefreedictionary.com/magnetite.

[121] Kirschvink, Joseph L., Atsuko Kobayashi-Kirschvink, Juan C. Diaz-Ricci, and Steven J. Kirschvink. "Magnetite in Human Tissues: A Mechanism for the Biological Effects of Weak ELF Magnetic Fields." Bioelectromagnetics Supplement 1 :101-113 (1992). 1992. Accessed February 22, 2019. http://web.gps.caltech.edu/~jkirschvink/pdfs/KirschvinkBEMS92.pdf.

[122] Gieré, Reto. "Magnetite in the Human Body: Biogenic vs. Anthropogenic." PNAS. October 25, 2016. Accessed February 22, 2019. doi:10.1073/pnas.1613349113.

[123] "Definition of Biomagnetism." The Free Dictionary. Accessed April 12, 2018. https://medical-dictionary.thefreedictionary.com/biomagnetism.

[124] Dhar, Michael. "How Hot Is the sun?" LiveScience. January 15, 2014. Accessed April 17, 2018. https://www.livescience.com/42593-how-hot-is-the-sun.html.

[125] Zell, Holly. "Hidden Portals in Earth's Magnetic Field." NASA. March 25, 2015. Accessed April 16, 2018. https://www.nasa.gov/mission_pages/sunearth/news/mag-portals.html.

[126] Ibid.

[127] Garner, Rob. "MMS Mission Overview." NASA. April 02, 2015. Accessed April 16, 2018. https://www.nasa.gov/mission_pages/mms/overview/index.html.

[128] Ibid.

[129] Ibid.

[130] Greenwald, John. "PPPL and Max Planck Physicists Reveal Experimental Verification of a Key Source of Fast Reconnection of Magnetic Fields." Princeton Plasma Physics Lab. March 31, 2017. Accessed April 16, 2018. https://www.pppl.gov/news/press-releases/2017/03/pppl-and-max-planck-physicists-reveal-experimental-verification-key.

[131] Rosen, Raphael. "Physicists Uncover Clues to Mechanism behind Magnetic Reconnection." Phys.org. January 23, 2017. Accessed April 16, 2018. https://phys.org/news/2017-01-physicists-uncover-clues-mechanism-magnetic.html#nRlv.

[132] Garner, Rob. "MMS Mission Overview." NASA. April 02, 2015. Accessed April 16, 2018. https://www.nasa.gov/mission_pages/mms/overview/index.html.

Chapter 15

A Slice or Two of Science on the Side

"Science never solves a problem without creating ten more."

— George Bernard Shaw

Science has repeatedly proven that disturbances of Earth's magnetosphere creates instabilities in technology and all living organisms, as well as on and in Earth itself.[133, 134, 135, 136, 137] HeartMath's lead researcher Rollin McCraty and Annette Deyhle wrote, "of all the bodily systems studied thus far, changes in geomagnetic conditions appear to most strongly affect the rhythms of the heart and brain."[138] This supposition is supported by researchers Halberg, et al,[139-140] Hamer,[141] Oraevskii, et al,[142] Pobachenko, et al,[143] Persinger, et al,[144-145] Otsuka, et al,[146] Dimitrova, et al,[147] and Rapoport, et al.[148]

Can Sasquatch Detect the Human Body's Magnetic Field?

McCraty and Deyhle also wrote, "… the heart's electromagnetic field can be detected by nearby animals or the nervous systems of other people."[149] They likewise inform that human bioemotional energy has a scientifically measurable nonlocal effect on people, events, and organic matter.[150] Finally, they claim, "… a bioelectromagnetic field such as the ones radiated by each human heart and brain can affect other individuals and the global information field environment."[151] Shocking! And quite pertinent to my sasquatch research.

Okay. Time for a short explanatory break. The above information—readily confirmed by many—is simply telling us that the human heart and brain have scientifically measurable electromagnetic fields, and that these electromagnetic fields can be detected by and affect not only others around us, but events and organic matter as well. Say what?

Wait a minute! Did I write that right? Did you read that right? Is science *really* telling us that the heart's measurable electromagnetic field can trigger a *change* in living organisms around us, and in things that happen around us? Can it also alter organic matter? Answers to these four questions are: Yes … Yes … Yes … Yes!

I suspect it is challenging to slip this idea comfortably within the realm of acceptance if you are wholly new to it. I get that. I assure you, however, HeartMath's two-plus decades of validated scientific research in this area (boring as much of it can be at times to read and digest) readily corroborates that startling proclamation. Let's take a bit of a breath, maybe have a brew—or a toke if you're the iconic Willy Nelson—then jump back into some of the mind-bending details.

Interestingly, of all the organs in the human body, the heart generates the "largest rhythmic electromagnetic field."[152] In fact, the electromagnetic field of the human heart is nearly 100 times stronger than that produced by the human brain.[153] Staggering, eh? Furthermore, the heart's field can be detected quite a few feet from the body using sensitive instrumentation. As mentioned earlier in this book, detection of it occurs at least 10 feet away from the body. Some suggest that the distance for detection of it is a bit farther than 10 feet; regardless, the heart's electromagnetic field extends outward from the physical body, is measurable, and, to reiterate, can be detected by and affect not only others around us, but events and organic matter as well.

Stated simply, this means that any time you are within about 10 feet or so of another person, the electromagnetic energy emanating out from each heart undergoes an energetic interaction—whether we want it to or not. Amazing! Puts a whole new twist to the idea of family dinners, meetings, congregations, concerts, professional sports games, shopping, protests, political gatherings, mob-rule, and so forth, eh? No wonder I dislike hanging around large crowds! Seriously.

Ponder on that idea for moment or so, then imagine that the same type of electromagnetic interactions could very well be true for sasquatches. That is, sasquatches are carbon-based beings and they have hearts, brains, and a nervous system. Thus, it stands to reason that they too emanate an electromagnetic field detectable through measurement, are *able to detect heart energy from others*, and also *affect others* when near them! Oh, my!

I don't know how you feel about this idea; however, I assure you that learning about and applying this very premise into my sasquatch research routine has produced unbelievable results. To wit, the latter part of my 2017 research when eight sasquatches visited my campsite as noted in Chapter 13.

If we wish to measure sasquatches emanating electromagnetic fields, all we need do is have the proper instrumentation available when within, say, 10-30 feet of them. No problem, eh? On the other hand, if we knew the science of coherence as put forth by the HeartMath Institute, learned how to apply it in any setting, and used it while interacting with sasquatches, what do think might happen? Well, for one thing, I believe you would be able to repeat some of the experiences I described in Chapters 12-13!

You read that correctly. I am shamelessly proposing:

Proposal 15:

> Sasquatch researchers who learn the act of coherence can use directed intent to control their interaction with sasquatches.

Learning to intentionally *control* and *direct* emanations of your electromagnetic heart field means that you can control the *type* and *intent* of the energy sasquatches detect. I'm convinced this is precisely how I ended up being able to attract and interact with the eight sasquatches who visited my campsite at the tail end of my 2017 research trip.

Furthermore, I believe sasquatches can *manipulate their emanated heart energy* in ways that may parallel what we are able to do using controlled and directed coherence. Of course, it isn't likely that sasquatches call it coherence. Even so, yet another astounding premise!

Bottom line: perhaps a different door to sasquatch research is opening. If so, the idea of heart/head *coherence* becomes not only workable, but also extremely relevant to this story of sasquatch and me. Hence, I offer here another proposal:

Proposal 16:

> Sasquatches—hairy, human-like apes—must be included in any discussion about biological systems involving electromagnetic energies.

Now then, on to a tad more science. You might think, "Yuk!" However, 'tis necessary for sure! For instance, please read the next quote and ask yourself if the idea is interesting or relevant—directly or obliquely—to this story. I, for one, believe it is both directly and obliquely interesting and relevant.

The following quote came from an online bowhunting magazine. An article titled, "Are Deer Detecting Your Electromagnetic Field?" was published on Jan. 13, 2011. In it, Jason Herbert, a hunter, author, and researcher, wrote, "Every living animal generates electromagnetic energy… *Research suggests animals can sense the electromagnetic field* (EMF) *produced by other animals* (author's emphasis).[154]

Herbert wrote the article with an idea in mind that he wanted to explore. That is, someone had developed EMF-blocking clothing to aid hunters in bagging their game more easily by allowing them to get closer to their game without detection. HECS, LLC, makes and sells this type of hunting clothing made up of a carbon fiber grid embedded within fabric; it is called Stealthscreen.[155] I read some of HECS research and it was both interesting and in keeping with many of the premises related to magnetic research. Check endnote reference number 154 if you wish to peruse their website. In the end, Herbert believes this clothing does, in fact, aid hunters to bag more game than otherwise. To each their own, eh?

I am not a hunter, although I hunted when I was a youngster; still, I found this information thought-provoking and cumulatively supportive of the idea that sasquatches emanate ELF electromagnetic frequencies. So, the question here becomes: "If you dropped a small fortune on energy-blocking clothing, could you get closer to sasquatches?" Maybe. What would you do if you could? Only you can answer that question. I believe, though, that wearing EMF blocking clothing while researching sasquatches would either inhibit or prohibit one's ability to intentionally engage in directed coherence. Always a tradeoff, eh?

Again, EMF blocking clothing is simply another small piece to a much larger jigsaw puzzle. Not only that, the idea of energy-blocking clothing is a tad funky, eh? Now then, to continue with science.

As mentioned earlier, literally all biological systems on Earth are touched daily by magnetic fields of varying strengths. Each magnetic field frequency interacts with, hence, alters in some manner, every cell within a biological system. In fact, it is now commonly known that many biological systems are affected during geomagnetic activity.[156] This, then, leads to another proposal:

Proposal 17:

> It's possible for sasquatches to manipulate electromagnetic energies to intentionally affect human autonomic nervous systems, brains, and cardiovascular systems, for purposes of self-defense or other reasons.

Continuing, in a study published in the February 2018 edition of *Scientific Reports*, the researchers wrote, "… findings support the [idea] that energetic environmental phenomena affect psychophysical [mental perception of physical stimuli] processes that can affect people in different ways depending on their sensitivity, health status and *capacity for self-regulation* (author's emphasis)."[157] Guess what? The idea of coherence includes these very same ideas! Please do keep this thought in mind, for after a bit or so, I'll be fleshing out the idea of *coherence* in greater depth.

Finally, it's important to realize and accept that Earth and the ionosphere work synchronously when producing resonant frequencies that, in the end, directly interact with the human brain and cardiovascular system. Even more central is the idea that *changes* in these frequencies affect living organisms' entire biological systems.[158] This is a significant awareness, worthy of added attention as we delve into the realm of how *intentional* use of coherence during my interactions with sasquatches is key to this story. Moreover, intentional, directed coherence contributed to many successful encounters with sasquatches throughout my 2017 research trip.

Let's say we find a way to bring *intentional* coherence between our physical heart/head connections, and that the frequencies created by this coherence either match or are compatible with the resonant frequencies sasquatches use while engaging in seeming paranormal activities. I offer that, if changes in coherence frequencies affect or influence the human autonomic nervous system, brain, and cardiovascular system, it becomes a near certainty sasquatches will respond to these projected frequencies in kind.

Hopefully, you are following the roadmap of electromagnetic energies I'm laying out. If you are, then you understand why I offered proposal 17.

To clarify, that method of influence could readily account for the way certain members of the sasquatch community claim to have been adversely affected by sasquatches. Examples of that include the way many folks report experiencing dizziness, headaches, or lack of mental clarity when near sasquatches, while others tell tales about experiencing fear, unrest, anger, general angst, or other forms of adverse emotional responses to sasquatches.

The act of coherence—a scientifically confirmed, measurable, and repeatable event—is an act that humans can learn to control, direct, and/or otherwise manipulate. I believe it is reasonable to presume that sasquatches may do this as well.

Hmm! Who the heck taught sasquatches' energetic manipulation? Maybe no one; such an act might very well be an innate one for them, embedded deeply within their DNA, much as a natural ability to detect Earth's magnetic fields for directional purposes could be. Obviously, I'm unable to produce empirical data to support this particular view right now. Of course, I have not seen empirical evidence to the contrary either. We simply do not know yet.

Next up, disturbances in Earth's ionosphere are based on premises of radiation, and we need to discuss two types of radiation found within the electromagnetic spectrum: ionizing and non-ionizing. We'll examine these next because the premise of radiation is a vital factor when engaging in intentional coherence.

Recall that coherence is a validated event involving a synchronous activity between the heart and brain. Also, heart energy radiates outward from the body at least 10 feet.

It is well known that we are continuously exposed to a variety of radiated energies which influence our ability to experience and express intentional coherence. Furthermore, the idea of ionizing radiation dovetails nicely with sasquatch researcher Pearl Prihoda's theory, which I address in greater depth in Chapter 16.

I'll first discuss ionizing radiation and then non-ionizing radiation. Following that, I'll elaborate on how these differing radiations fit into this story of sasquatch and me, coherence, and energy manipulation.

Ionizing Radiation

According to Ali Zamanian and Cy Hardiman, researchers who study the effects of electromagnetic radiation on human health, "*Ionizing radiation* contains sufficient electromagnetic energy to strip atoms and molecules from the tissue and alter chemical reactions in the body (converting molecules totally or partly into ions)."[159] Dang! This doesn't sound too healthy, does it? They also wrote, "Human beings are constantly exposed to low levels of ionizing radiation from natural sources." [160] The type of radiation discussed here is known as *natural background* radiation. Primary sources for natural background radiation include visible light; ultraviolet and infrared light (from the sun);

radioactive materials on Earth's surface, such as coal and granite, radon, cosmic rays, and natural radioactivity in the human body.[161]

Interestingly, granite countertops also emit radiation. The U.S. Environmental Protection Agency (EPA) and the U.S. Nuclear Regulatory Commission (NRC) have created standards for acceptable radiation from granite countertops. In most instances, granite countertops do not present a danger. If an excessive amount of radon emission were measured from granite countertops, however, this would be due to the natural uranium and/or thorium concentration found in the granite.[162-163]

Okay, we are bombarded with a bunch of natural background ionizing radiation every day of our lives. Too, I'd like to remind you that the Laurentian Shield is an "underlying formation of old, dry granite."[164] This is a prominent reason why the Laurentian Shield is a central factor to this story.

Stated simply, "ionizing radiation is any type of particle or electromagnetic wave that carries enough energy to ionize or remove electrons from an atom."[165] That is, ionizing radiation strips electrons from atoms. Remarkably, this idea is a key part of sasquatch researcher Pearl Prihoda's PPIT theory discussed in Chapter 16.

A stimulating and pertinent detail about ionizing radiation relates to *potassium*, which is another key ingredient for Prihoda's PPIT theory. To start, "we require potassium to survive."[166] Potassium is found naturally within nature and is a required mineral for the human body. The human body "needs potassium to build proteins, break down and use carbohydrates, build muscle, maintain normal body growth, *control the electrical activity of the heart*, and control the acid-base balance within the body (author's emphasis)."[167]

Inasmuch as coherence consists of intentionally synchronizing energy between the heart and head—including heart-rate variability (a variation in time intervals between heart beats, which is essentially an electrical signal)—I suspect an individual with a low potassium count might have a harder time moving into coherence than one with a normal count of potassium. Food for thought? Keep in mind, please, a synchronous, harmonious connection between the heart and head demands that we pay a bit of attention to details of our health—something I confess I have not always done well.

We can readily define potassium as "... one of the main blood minerals called electrolytes ... which means it *carries a tiny electrical charge* [or potential] (author's emphasis)."[168] There you have it! Potassium is yet another of the cumulative pieces of a whole, and it carries a tiny electrical charge. No wonder Prihoda's PPIT theory includes it. Next up, let's look at non-ionizing radiation and see if we can figure out how it fits into this story.

Non-Ionizing Radiation

The U. S. Department of Labor defines non-ionizing radiation as "a series of energy waves composed of oscillating electric and magnetic fields traveling at the speed of light. Non-ionizing radiation includes the spectrum of ultraviolet (UV), visible light, infrared (IR), microwave (MW), radio frequency (RF), and extremely low frequency (ELF)."[169] From the previous, the idea of oscillating electric and magnetic fields are important to our theory, as are UV, IR, and ELF.

The National Institutes of Health (NIH), Division of Occupational Health and Safety, states, "[Non-ionizing Radiation] lacks sufficient energy to remove electrons from atoms and molecules and is generally perceived as less harmful [than ionizing radiation]."[170] Ah hah!

This then, is the major distinction between ionizing and non-ionizing radiation. Ionizing radiation plays *whack-an-atom* by randomly knocking electrons off atoms, while non-ionizing radiation does not. Sounds awfully much like the old arcade game, *Whac-a-Mole*, where plastic moles popped up randomly through holes and the player had to hit the mole with a large, soft mallet within a specific timeframe. In this instance, ionizing radiation serves as the hammer to electrons.

Our primary interest with non-ionizing radiation lies in the fact that it is a series of energy waves composed of oscillating electric and magnetic fields traveling at the speed of light. The Centers for Disease Control (CDC) informs us, "Put simply, non-ionizing radiation differs from ionizing radiation in the way it acts on materials like air, water, and living tissue."[171] They also state, "non-ionizing radiation can heat substances."[172]

You read that right, and we have a bit of interest in the latter fact. Why? Because IR is classified as non-ionizing radiation; it produces *heat*, and we have shown that IR influences sasquatch encounters. Remember, those alert, elusive sasquatches seem perfectly capable of detecting heat signatures of IR camera traps. Clever creatures, eh? So, sasquatches have a built-in thermographic ability? Now that's just silly! — Or is it?

We now know a little bit about how ionizing and non-ionizing radiation works and the primary difference between them. We also know that IR fits under the umbrella of non-ionizing radiation. And, as frequencies would have it, ELF is likewise a form of non-ionizing radiation. Further, very low frequency (VLF) radio propagations also occur naturally in the environment and they, too, are non-ionizing radiation.

Additionally, certain noises contribute to ELF and VLF frequencies. For example, lightning discharges during thunderstorms, volcanic eruptions, dust storms, and tornadoes are all known to serve as natural origins for both ELF and VLF. Believe it or not, each of these can also be tied directly to space weather. Furthermore, they each affect biological systems. Fascinating! The evidence continues to pile up!

Human-made sources are also known to create ELF and VLF radio propagation waves. For example, you may recall from Chapter 2 our discussion about the ELF antennas used by the U. S. Navy to communicate with submarines around the world. A seriously large antenna was needed for that to occur. Recall also, that the Navy used the natural formation of Earth via the Laurentian Shield to functionally increase their ELF antenna range. Why? Because the makeup of the Laurentian Shield is a *natural* way for ELF signals to travel through Earth. Pertinent!

When the Navy's ELF project was in full bloom, myriad long-wave non-ionizing radiation traveled through the Laurentian Shield. Movement of those waves, then, affected not only our early magnetic heart-head research results, it provided sasquatches with readily accessed energies with which to perform their seeming magic.

Allow me to expand a bit on the way electrical storms affect electromagnetism. To do this, I'll introduce another odd term: *sferic*. Pierce, a NASA technical report writer, wrote:

> A radio atmospheric signal or *sferic* is a broadband electromagnetic impulse that occurs as a result of natural atmospheric lightning discharges. Sferics is defined most commonly as 'atmospheric interference'; the actual properties of sferics are '*the electric and magnetic fields generated by electrified clouds and lightning flashes*' (author's emphasis).[173]

Wikipedia informs us that "sferics may propagate from their lightning source without major attenuation in the earth-ionosphere waveguide, and can be received thousands of kilometers from their source."[174] The idea that sferics can travel thousands of miles (kilometers) from their source is pertinent here, because the frequencies involved are the very same ones causing alterations to, or otherwise influencing human autonomic nervous systems, brains, and cardiovascular systems. Moreover, might this known anomaly contribute to the reason that sasquatches interact with me more frequently during electrical storms? I believe this is possible!

Again, though, the question arises, "Is it possible that sasquatches can manipulate some of these energies when they occur?" I lean strongly toward yes.

Finally, and this, too, is quite significant, sferics frequencies exist in the lower ELF band and they seem coincident with something known as Schumann Resonances (SR). That is, the waveband within sferics is typically considered as 3-60 Hz, while SR frequencies propagate between 7.83 Hz to around 45 Hz, which means they fall well into the band of frequencies known as ELF.

We certainly now know that ELF plays a role in human behavior; I'm suggesting the same is true for sasquatches. To further probe this idea, I next look at how Schumann Resonances are defined. As well, I address how they might play a role in sasquatch behavior.

Schumann Resonances

Schumann Resonances form in what is known as the conductive ionosphere. The conductive ionosphere is known to act as a wave-guide. A waveguide "... confine[s] and direct[s] the propagation of electromagnetic waves such as light."[175] According to HeartMath's Annette Deyhle, "The limited dimensions of the earth cause this [ionospheric] wave-guide to act as a resonant cavity for electromagnetic (EM) waves in what are extremely low frequencies bands."[176]

Once again, the fundamental frequency within the SR is 7.83 Hz. The other frequencies are known generally as 14, 20, 26, 33, and 45 Hz; hence, we know SRs lie well within the ELF band, which is a significant band of frequencies to our story.

NASA wrote of Schumann Resonances:

> At any given moment about 2,000 thunderstorms roll over Earth, producing some 50 flashes of lightning every second. Each lightning burst creates electromagnetic waves that begin to circle around Earth captured between Earth's surface and a boundary about 60 miles up. Some of the waves—if they have just the right wavelength—combine, increasing in strength, to create a repeating atmospheric heartbeat known as Schumann Resonance.[177]

In addition, the waves of energy mentioned above are extremely *long waves*—nearly the length of Earth's circumference. We already know that ELF waves propagate as *long waves*; thus, this information appears to confirm another clear association between SRs and ELF. That is, SR long waves contribute to sasquatches' abilities to manipulate energy.

Schumann Resonances became a validated part of science when they were measured consistently in the early 1960s. Much has happened since then that adds information to our knowledgebase of SRs. For example:

> Scientists have discovered that variations in the [Schumann] resonances correspond to changes in the seasons, solar activity, activity in Earth's magnetic environment, in water aerosols in the atmosphere, and other Earth-bound phenomena.[178]

This is not only fascinating information, it surely supports earlier premises about the importance of ELF to our story.

Nikola Tesla, a Serbian-American inventor and futurist who is best known for his contributions to the design of modern alternating current (AC) electricity supply system, referred to the importance of SR, and was quoted in a narrated online film as saying:

> Alpha waves in the human brain are between 6 and 8 hertz. The wave frequency of the human cavity resonates between 6 and 8 hertz. All biological systems operate in the same frequency range. The human brain's alpha waves function in this range and the electrical resonance of the earth [Schumann Resonances] is between 6 and 8 hertz. Thus, *our entire biological system—the brain and the earth itself—work on the same frequencies* (author's emphasis).[179]

Please take due note of this quote. It reflects the importance of understanding a bit about the electromagnetic qualities of the human brain, as well as the way these qualities interact with Earth's SR's and ELF. Why? Because we've proved an incredibly relevant fact: Our entire biological system—the brain and the earth itself—work on the same frequencies! This is applicable here because I am striving to show how ELFs naturally affect not only us, but those elusive sasquatches about whom we so dearly want to learn.

The actual end of Tesla's quote might cause a bit of consternation and even a smidgen of controversy for some; however, I believe I'll include it anyway. Tesla ended his above quote with, "If we can control that resonate system electronically, we can directly control the entire mental system of humankind."[180] See what I mean? Yikes!

Tesla made that statement in the early 20th century; way before electronics evolved into the high-tech, miniaturized, nanotechnology, that exists today. My point in including that part of the quote has nothing to do with potential ramifications surrounding possible actualization of such a theory. Rather, it has to do with Tesla's clear awareness of the way in which SRs affect the human brain. And once again, if human brains are affected, we must surely accept that the brains of sasquatches are as well. Or so I believe.

Plus, it seems to me that if sasquatches are somehow able to manipulate these energies, important, previously unanswered questions within the sasquatch community might be nearing at least possible—if not probable—answers. Open minds often lead to credible solutions. Please stay open-minded.

It is well known that a clear and ongoing link exists between the ionospheric cavity, the human brain, and the human nervous system. It seems both fair and reasonable to transfer that same link to the brains and nervous systems of sasquatches. Unless, of course, sasquatches are ultimately proven to be totally alien, which seems highly unlikely to me.

Based on personal experience, they are carbon-based entities with similar organic attributes as humans. Irrespective, that which affects humans must also affect sasquatches; seems like common sense to me. Many, however, may readily refute this thought; so be it. As my personal research continues, I hope one day to offer empirical data to support all my proposals. Yes … I do indeed hold high hopes!

Next, I address the concept of plasma as we try to understand how sasquatches make use of ELF to induce a so-called plasma effect often found in photos or videos when we use IR technology. Again, it is common knowledge within the sasquatch community that many so-called blob photos have circular representations of what many believe is plasma. In turn, these circular spots tend to disguise or cover up sasquatches in photos or videos. You saw some of these circular spots in photos that I posted in Chapter 13. Our goal here is to gain insight into how this might occur and, to get a sense for how plasma plays a role in Prihoda's PPIT hypothesis.

A Bit About Plasma

If we wish to discuss how sasquatches might use plasma to influence IR photos and videos, we need to be fully aware of what plasma is and its origin—which is the sun. We also must know something about its path from the sun to Earth.

Princeton's Plasma Physics Laboratory wrote:

> The explosive force of solar flares can be equivalent to millions of tons of TNT. Plasma, the medium in which reconnection takes place, makes up the sun and stars and more than 99 percent of the visible universe.[181]

Wow! That's bunches of plasma assailing our bodies every single minute, hour, day, week, month, and year! Furthermore, NASA informs us that plasma is a fourth state of matter and it exists in space. In that state, atoms are positively charged and share space with free negatively charged electrons. In the end, it turns out that the solar wind is hot plasma blowing from the sun.[182]

Even though it's important to recognize that the solar wind is hot plasma, our interest here lies predominantly with NASA's next statement. That is, plasma has the ability "to conduct electricity and interact strongly with electric and magnetic fields."[183] Seems like almost everything in this story of sasquatch and me relates to the ideas of electricity and magnetic fields! Indeed! That is why I spend so much time discussing them. Plus, the electromagnetic qualities of plasma play a role in Prihoda's PPIT theory examined in Chapter 16.

Plasma leaves the sun during solar flares, filament eruptions, and coronal mass ejections (CMEs). Plasma particles ejected off the sun are ridiculously hot and travel at enormously high speeds. For example, NASA informs us that, "solar plasma is heated to tens of millions of degrees, and electrons, protons, and heavy nuclei are accelerated to near the speed of light."[184] Hmm … so, plasma is both hot and fast. Sounds like some of our grandchildren!

When directed towards Earth, these fast-moving plasma particles first meet Earth's magnetosphere—that protective bubble of energy surrounding Earth. You may recall a key role of the magnetosphere is to "control the motion of charged particles,"[185] in part by deflecting many of them from Earth, thereby protecting Earth from the harmful effects they produce.

Finally, I suggest that the electromagnetic qualities of plasma serve as the culprit when snapping photos of sasquatches or shooting videos of them when using devices or camera traps employing IR. This discussion leads naturally to another proposal:

Proposal 18:

> Sasquatches can manipulate plasma, thereby causing distortions in photos and videos.

Sure … and this sounds a bit off, eh? Perhaps. My sasquatch proposals grant them abilities to detect and manipulate energy, so allowing them added leeway for this act doesn't seem too big a stretch to me. All we really need is more time, funds for proper instrumentation, and perseverance to set up proof for such claims.

Next on deck is a peek at photosynthesis, which similarly plays a role in Prihoda's PPIT theory. After all, the full name of her theory is Photosynthetic Piezoelectric Induced Transparency.

Photosynthesis

Most folks likely remember from middle school days that photosynthesis refers to the process "by which green plants and certain other organisms transform light energy into chemical energy."[186] Chemically speaking:

> … photosynthesis is a light-energized oxidation–reduction process. (Oxidation refers to the removal of electrons from a molecule; reduction refers to the gain of electrons by a molecule.)[187]

The simple fact of a light-energized oxidation-reduction process makes photosynthesis pertinent to Prihoda's PPIT theory. Also, a few distinct minerals are needed for plants to be healthy, thereby experiencing the greatest rates of photosynthesis. The following minerals are included in this requirement: nitrogen, sulfate, phosphate, iron, magnesium, calcium, and potassium. Prihoda's theory addresses several of these minerals. Also, molecules of chlorophyll serve as the primary pigment in all green plants. These molecules "are arranged within a leaf such that they minimize the plant's need to transport incoming solar radiation while also increasing a leaf's photosynthetic output."[188]

Let me remind you of another key point about photosynthesis: "… when light gets to a plan, the plant doesn't use all of it." [189] In fact, "…only certain *colors* [are used] to make photosynthesis happen." [190] Actually, "plants mostly absorb **red** and **blue** wavelengths." [191] This fact is relevant because the colors red and blue are part of the visible light spectrum, which, in turn, is part of the overall electromagnetic spectrum.

Furthermore, infrared frequencies fall just below the *red* wavelength, so it's easy to see how close red is to infrared. As light waves, each color has a typical wavelength and a distinctive frequency. This book is not the forum for detailed discussion on color wavelengths and frequencies, although I do offer a brief introduction to the idea in Chapter 16.

Next, I'll step cheekily out on another thin limb. Many witnesses have reported a red or slightly yellowish glow coming from the eyes of sasquatches. This, then, opens a door to an additional proposal.

Proposal 19:

> Sasquatch eye glowing incidents link with energetic waveforms connecting infrared frequencies and magnetoreception qualities as they interact with Earth's changing magnetic field.

Next on deck is a discussion on coherence. By the end of the chapter you should be well-apprised of why I believe that coherence is a relevant research tool.

Coherence

Coherence is defined typically and scientifically. Common dictionary definitions for coherence refer to a system wide order, be the system prose, art, psychological, physical, etc. In physics, coherence refers primarily to waveform states. I wish, however, to focus on the typical definition in this context rather than the scientific or physics one.

To begin, the HeartMath Institute has spent decades researching the premise of head/heart coherence. What I've learned about the concept of coherence came largely from my studies of the HeartMath materials combined with my earlier heart/head research. Becoming proficient in the act of coherence occurs through a combination of education and diligent practice, although I suggest that belief and perceptions also play a role.

HeartMath uses the term coherence as an "umbrella term to describe a physiological mode that encompasses entrainment, resonance, and synchronization—distinct but related phenomena, all of which emerge from the harmonious activity and interactions of the body's subsystems."[192] The previous bespeaks the necessity for including waveform states into the discussion. Further, head/heart coherence helps the human system in several ways, including creation of "system-wide energy efficiency."[193] The latter quote addresses the need to examine the typical definition for coherence. As happenstance would have it, *system-wide energy efficiency* is the central factor in how I employ the act of coherence while interacting with sasquatches. This is an important point to keep in mind.

Basically, using a tried and proven coherence technique serves the purpose of aligning the body's physical system such that an intentional and controlled emanation of so-called heart energy is possible. By now you are quite familiar with the fact that the heart emanates an electromagnetic field. I've persistently disclosed this, and Chapter 13 spelled out incredible results of such activity.

Although the science underlying coherence as posited by HeartMath is lengthy and quite complex at times, learning how to engage in coherence is not. Discussing the in-depth science of coherence falls outside the scope of this book; however, many of the endnotes give references you may pursue if you are interested in discovering more about it. One aspect of coherence, though, *is* important enough to spend a bit of time discussing: heart-rate variability (HRV).

Heart Rate Variability

Time exists between each beat of the heart, and HRV is simply a measure in the difference of that time. An interesting and powerful feature of HRV relates to the way science has authenticated direct associations between HRV, physical processes within the human body, mental clarity, and emotions.

HRV, used as a measure for system-wide energy efficiency, has become a standard within the scientific and medical communities. In fact, many organizations have developed non-invasive, relatively low-cost devices that combine measurement of physical parameters along with internal standard calculations to readily determine HRV.

Various techniques and non-invasive devices that aid one to reach a desired state of system-wide energy efficiency by learning to control HRV are on the market. I am not interested in marketing specific devices or techniques; even so, I admit to bias about HeartMath Institute and their development of certain products. Their techniques and products hit the marketplace early, even before the concept of HRV had captured worldwide interest. These products have improved greatly over the years and their acceptance by many professionals within medical, corporate, and educational communities soundly substantiate their validity and usefulness.

If you are serious about learning to control your personal HRV in an attempt to reach and sustain a sense of system-wide energy efficiency, I highly recommend checking out HeartMath's products:

1. *Inner Balance*, a Bluetooth device that works with iPhone, iPad, and Android smartphones.
2. *emWave2*, a standalone unit that connects to a computer, laptop, etc.
3. *emWave Pro*, a computer-based HRV detector/assessment method used by many professionals in various health care and mental health settings, as well as by many corporate entities to aid employees to attain and sustain greater personal health and productivity.

Finally, learning to control one's HRV is central to reaching and sustaining intentional and directed coherence, or, if you will, a system-wide energy efficiency used to productively engage with sasquatches. I personally employ the following technique while interacting with sasquatches while wandering the vast wildernesses in which they live and thrive.

System-wide Energy Efficiency Exercise

The following exercise is adapted from HeartMath's Introductory Coherence Technique.[194] I suspect folks familiar with meditation techniques will be conversant with some of the wording used in this system. Point of note, using this technique has allowed Julie and me to enjoy many pleasant experiences with sasquatches … repeatable experiences.

Exercise Steps

1. Bring yourself to a state of calm using whatever method works best for you; for example, breathing exercises are helpful.
2. Focus on the dynamic qualities associated with **appreciation**; that is, appreciate your wilderness surroundings and express gratitude for being able to interact with sasquatches and all living organisms within the wilderness.
3. Allow your focus to include a sense of **compassion** for all around you; feel empathy for all living organisms in the wilderness, specifically, sasquatches.
4. Turn your attention to your heart and imagine your sense of appreciation, forgiveness, understanding, compassion, valor, and humility flowing outward from your heart into your surroundings. This is key because I'm convinced sasquatches can detect this outward flow of energy, even if they are miles distant from me. My experience with this step also includes nearby sasquatches engaging in vocalizations as I directed this energy outward, almost as if they were surprised by this intentional energetic radiation and were striving to figure out who I am or what I'm doing.
5. Imagine this radiated energy flowing farther and farther outward into the surrounding wilderness (if that is where you are). Remember, we have proven that ELF waveforms travel long distances, and system-wide energy efficiency (coherence) radiates such waveforms. If sasquatches are not nearby, the traveling waveforms radiated outward are bound to attract them if you are engaging in a sincere attempt to interact with them. For those who might think this technique could be used to entice sasquatches to come close enough to do them harm, I assure you that the sasquatches I've interacted with while using this technique were able to detect heartfelt genuineness. I strongly suspect they are as adept at detecting either a lack of sincerity or an intent to cause them harm.

6. Envision sasquatches responding to this energy in a pleasant, beneficial way.
7. Finally, enjoy and document your results for future evaluation.

That's it! No big deal. Simple, quick, and certainly profoundly effective in my research. Be sure not to allow cognitive dissonance or cognitive disruption of any type prevent you from recording any anomalous experiences you might have while engaging in this exercise.

Finally, be aware that this technique may not work the first time you try it. No worries; simply practice … practice … practice.

Chapter 15

[133] Garner, Rob. "MMS Mission Overview." NASA. April 02, 2015. Accessed April 16, 2018. https://www.nasa.gov/mission_pages/mms/overview/index.html.

[133] Staff, NOAA. "Space Weather Impacts on Climate." Space Weather Impacts On Climate. Accessed April 12, 2018. https://www.bing.com/cr?IG=14E9A709695E492B9B1EE4746FEA1748&CID=2F21EC1263256C5F2396E7DD628A6DB0&rd=1&h=_62xhUTLUvVVwpNB-F7TnxfHnwE2kzc9iD-FhDFR66E&v=1&r=https%3a%2f%2fwww.swpc.noaa.gov%2fimpacts%2fspace-weather-impacts-climate&p=DevEx,5069.1.

[134] Dunbar, Brian. "How Space Weather Affects Space Exploration." Van Allen Probes. Accessed April 12, 2018. https://www.nasa.gov/mission_pages/rbsp/science/rbsp-spaceweather-human.html.

[135] Chu, Jennifer. "Space Weather's Effects on Satellites." MIT News. Accessed April 12, 2018. https://news.mit.edu/2013/space-weather-effects-on-satellites-0917.

[136] Russell, Randy. "How Does Space Weather Affect Earth and Human Society?" Windows to the Universe. Accessed April 12, 2018.

https://www.windows2universe.org/?page=%2Fspace_weather%2Fsw_intro%2Fsw_affect_us.html.

137 Alexander, Latlitha T., Naifa Suliman Al Atawi, and Hala Mohamed Abo Mostafa. "Space Weather Effects on Humans in Tabuk City, KSA." *International Journal of Applied Science and Technology Vol. 6, No. 1; February 2016* 6, no. 1 (February 2016): 47-57.

138 "The Global Coherence Initiative: Investigating the Dynamic Relationship Between People and Earth's Energetic Systems." HeartMath Institute. October 9, 2014. Accessed April 20, 2018. https://www.heartmath.org/research/research-library/coherence/global-coherence-initiative-investigating-dynamic-relationship-people-earths-energetic-systems/, 5.

139 Halberg, Franz & Cornélissen, Germaine & Otsuka, Kuniaki & Watanabe, Yoshihiko & Katinas, George & Burioka, Naoto & Delyukov, Anatoly & Gorgo, Yuriy & Zhao, Ziyan & Weydahl, Andi & Sothern, Robert & Siegelova, Jarmila & Fiser, Bohumil & Dusek, Jiri & V. Syutkina, Elena & Perfetto, Federico & Tarquini, Roberto & Singh, Ram & Rhees, Brad & International BIOCOS Study Group, the. (2000). Accessed April 20, 2018. Cross-spectrally coherent ~10.5- and 21-year biological and physical cycles, magnetic storms and myocardial infarctions*. Neuro endocrinology letters. 21. 233-258.

140 Halberg, Franz & Cornélissen, Germaine & Sothern, Robert & Katinas, George & Schwartzkopff, Othild & Otsuka, Kuniaki. (2008). Cycles Tipping the Scale between Death and Survival (=``Life. Progress of Theoretical Physics Supplement - PROG THEOR PHYS SUPPL. 173. 153-181. 10.1143/PTPS.173.153. Accessed April 20, 2018.

141 Hamer, J. R. "Biological Entrainment of the Human Brain by Low Frequency Radiation (NSL 65-199)." *Palos Verdes Peninsula, Calif.: Northrup Space'Labs* (1965): 17.

142 N Oraevskiĭ, V & Breus, Tamara & M Baevskiĭ, R & I Rapoport, S & M Petrov, V & V Barsukova, Zh & I Gurfinkel, Iu & Rogoza, Anatoly. (1998). Effect of geomagnetic activity on the functional status of the body. Biofizika. 43. 819-26. Accessed April 20, 2018.

[143] Pobachenko, S. V., A. G. Kolesnik, A. S. Borodin, and V. V. Kalyuzhin. "The Contingency of Parameters of Human Encephalograms and Schumann Resonance Electromagnetic Fields Revealed in Monitoring Studies." SpringerLink. March 3, 2005. Accessed April 20, 2018. doi:10.1134/S0006350906030225.

[144] Persinger, M. A., and C. Pscyh. "Sudden Unexpected Death in Epileptics following Sudden, Intense, Increases in Geomagnetic Activity: Prevalence of Effect and Potential Mechanisms." *International Journal of Biometeorology* 38, no. 4 (January 5, 1995): 180-87. doi:10.1007/BF01245386.

[145] Persinger, M. A. "Geopsychology and Geopsychopathology: Mental Processes and Disorders Associated with Geochemical and Geophysical Factors." *Experentia* 43, no. 1 (January 1987): 92.104. doi:10.1007/BF01940360. Accessed April 20, 2018.

[146] Otsuka, Kuniaki, Germaine Cornelissen, Tsering Norboo, Emiko Takasugi, and Franz Halberg. "Chronomics and "Glocal" (Combined Global and Local) Assessment of Human Life | Progress of Theoretical Physics Supplement | Oxford Academic." OUP Academic. February 01, 2008. Accessed April 20, 2018. doi:10.1143/PTPS.173.134.

[147] Dimitrova, S., I. Stoilova, and I. Cholokov. "Influence of Local Geomagnetic Storms on Arterial Blood Pressure." *Bio Electro Magnetics* 25, no. 6 (August 3, 2004): 408-14. Accessed April 20, 2018. doi:10.1002/bem.20009.

[148] Rapoport, I., N. K. Malinovskaia, V. N. Oraevskií, F. I. Komorov, A. M. Nosovskií, and L. Vetterberg. "Effects of Disturbances of Natural Magnetic Field of the earth on Melatonin Production in Patients with Coronary Heart Disease." ResearchGate. February 1997. Accessed April 20, 2018. https://www.researchgate.net/publication/13947869_Effects_of_disturbances_of_natural_magnetic_field_of_the_Earth_on_melatonin_production_in_patients_with_coronary_heart_disease. *Klinicheskaia Meditsina (Mosk)* 1997;75(6):24–6.

[149] "The Global Coherence Initiative: Investigating the Dynamic Relationship Between People and Earth's Energetic Systems." HeartMath Institute. October 9, 2014. Accessed April 20, 2018. https://www.heartmath.org/research/research-library/coherence/global-coherence-initiative-investigating-dynamic-relationship-people-earths-energetic-systems/, 9.

[150] "The Global Coherence Initiative: Investigating the Dynamic Relationship Between People and Earth's Energetic Systems." HeartMath Institute. October 9, 2014. Accessed April 20, 2018. https://www.heartmath.org/research/research-library/coherence/global-coherence-initiative-investigating-dynamic-relationship-people-earths-energetic-systems/.

[151] Ibid.

[152] Ibid.

[153] Ibid.

[154] Herbert, Jason. "Are Deer Detecting Your Electromagnetic Energy?" *Petersen's Bowhunting*, Petersen's Bowhunting Magazine, 13 Jan. 2011. Accessed 9 July 2018. www.bowhuntingmag.com/tactics/tactics_are_deer_detecting_your_electromagnetic_energy_011311/.

[155] HECS, LLC, Staff. "HECS STEALTHSCREEN." *Animals Sense EM Signal | HECS Stealthscreen*, HECS Stealthscreen. Accessed 9 July 2018. www.hecsllc.com/.

[156] McCraty, Rollin, Atkinson, Mike, Viktor Stolc, Abdullah A. Alabdulgader, Alfonsas Vainoras, and Minvydas Ragulskis. "Synchronization of Human Autonomic Nervous System Rhythms with Geomagnetic Activity in Human Subjects." MDPI. July 13, 2017. Accessed April 20, 2018. doi:10.3390/ijerph14070770, 1.

[157] Alabdulgader, Abdullah, Rollin McCraty, Michael Atkinson, York Dobyns, Alfonsas Vainoras, Minvydas Ragulskis, and Viktor Stolc. "Long-Term Study of Heart Rate Variability Responses to Changes in the Solar and Geomagnetic Environment." Nature News. February 08, 2018. Accessed April 21, 2018. doi:10.1038/s41598-018-20932.x. Scientific Reports is one of the journals published by Nature magazine. https://www.nature.com/srep/.

[158] McCraty, Rollin. *Science of the Heart: Exploring the Role of Heart in Human Performance.* Vol. 2. Boulder Creek: HeartMath Institute, 2015, 93.
Note: According to HeartMath Institute research, "The heart is, in fact a highly complex information-processing center with its own functional brain, commonly called the *heart-brain*, that communicates with and influences the

cranial brain via the nervous system, hormonal system and other pathways. These ... affect brain function and most of the body's major organs," 1-2.

[159] Zamanian, Ali, and Hardiman, Cy. "Electromagnetic Radiation and Human Health: A Review of Sources and Effects." *High Frequency Electronics*, July 2005, 17-26. Accessed June 18, 2018 from http://www.highfrequencyelectronics.com/Jul05/HFE0705_Zamanian.pdf, 18.

[160] Ibid., 18.

[161] Ibid., 18.

[162] Health Physics Society. "Radiation from Granite Countertops Fact Sheet." Health Physics Society - Specialists in Radiation Safety. Accessed January 09, 2019. https://hps.org/documents/Radiation_granite_countertops.pdf.

[163] Murphy, Kate. "What's Lurking in Your Countertop?" The New York Times. July 24, 2008. Accessed January 09, 2019. https://hps.org/documents/nyt_countertop_7-24-08.pdf.

[164] Biologic Effects of Electric and Magnetic Fields Associated with Proposed Project Seafarer. Ft. Belvoir: Defense Technical Information Center, 1977. doi:10.17226/20341.

Note: This report was prepared in 1977 by the Committee on Biosphere Effects of Extremely-Low-Frequency Radiation in response to a request from the United States Navy.

[165] Doss, H. M. "Ionizing Radiation and Humans – The Basics." PhysicsCentral. Accessed June 18, 2018. http://www.physicscentral.com/explore/action/radiationandhumans.cfm.

[166] Ibid.

[167] MedlinePlus, Staff. "Potassium in Diet: MedlinePlus Medical Encyclopedia." MedlinePlus. Accessed June 18, 2018. https://medlineplus.gov/ency/article/002413.htm.

[168] Haas, Elson M., M.D. "Role of Potassium in Maintaining Health." Periodic Paralysis International. July 17, 2011. Accessed June 18, 2018. http://hkpp.org/patients/potassium-health.

Note: The article was copyrighted in 2000 and placed online July 17, 2011.

[169] Department of Labor, Staff. "UNITED STATES DEPARTMENT OF LABOR." Occupational Safety and Health Administration. Accessed June 20, 2018. https://www.osha.gov/SLTC/radiation_nonionizing/index.html.

[170] NIH, Staff. "Non-Ionizing Radiation." National Institutes of Health. Accessed June 20, 2018. https://www.ors.od.nih.gov/sr/dohs/safety/Pages/non-ionizing.aspx.
[171] CDC, Staff. "Non-Ionizing Radiation." Centers for Disease Control and Prevention. December 07, 2015. Accessed June 20, 2018. https://www.cdc.gov/nceh/radiation/nonionizing_radiation.html.
[172] Ibid.
[173] Pierce, E. T. "Sferics." NTRS - NASA Technical Reports Server. November 22, 1995. Accessed June 22, 2018. https://ntrs.nasa.gov/search.jsp?R=19720017733.
[174] "Radio Atmospheric." Wikipedia. June 19, 2018. Accessed June 20, 2018. https://en.wikipedia.org/wiki/Radio_atmospheric.
[175] "Waveguide." Merriam-Webster. Accessed June 22, 2018. https://www.merriam-webster.com/dictionary/waveguide.
[176] Deyhle, Annette. "Influence of Geomagnetism and Schumann Resonances on Human Health and Behavior." HeartMath Institute. April 13, 2015. Accessed June 21, 2018. https://www.heartmath.org/gci-commentaries/influence-of-geomagnetism-and-schumann-resonances-on-human-health-and-behavior/, 2.
[177] Dunbar, Brian. "Schumann Resonance." NASA. Accessed July 26, 2018. https://www.nasa.gov/mission_pages/sunearth/news/gallery/schumann-resonance.html.
[178] Ibid.
[179] *Lost Lighting: The Missing Secrets of Nikola Tesla.* Produced by Steven R. Naft. Directed by Jay Miracle. By Jay Miracle and Scott JT Frank. Performed by Host: Dean Stockwell Narrator: Bill Rogers. DocumentaryTube.com. Accessed June 22, 2018. http://www.documentarytube.com/videos/lost-lightning-the-missing-secrets-of-nikola-tesla.
[180] Ibid.
[181] PPPL, Staff. "PPPL Scientists Bring Mysterious Process down to Earth." Fact Sheet. Princeton Plasma Physics Lab. October 2011. Accessed June 27, 2018. https://www.pppl.gov/news/2011/09/pppl-scientists-bring-mysterious-process-down-earth.

[182] NASA, Staff. "NASA Science Beta Glossary." NASA Science Beta. Accessed June 28, 2018. https://science.nasa.gov/glossary/p.

[183] Ibid.

[184] NASA, Staff. "Coronal Mass Ejections." NASA - Cosmicopia: An Abundance of Cosmic Rays. Accessed June 28, 2018. https://helios.gsfc.nasa.gov/cme.html.

[185] Ibid.

[186] Lambers, Hans, and James Alan Bassham. "Photosynthesis." Encyclopædia Britannica. June 15, 2018. Accessed June 28, 2018. https://www.britannica.com/science/photosynthesis.

[187] Ibid.

[188] Ibid.

[189] Rader, Andrew. "Photosynthesis - Part I: The sun and Light." *Biology4Kids.Com: Cell Structure: Cell Walls*, Andrew Rader Studios. Accessed 11 July 2018. www.biology4kids.com/files/plants_photosynthesis.html.

[190] Ibid.

[191] Ibid.

[192] McCraty, Rollin. "The Energetic Heart - GCI Edition." *HeartMath Institute*, GCI, 2003, www.heartmath.org/gci/resources/downloads/the-energetic-heart-gci-edition/, 4.

[192] Ibid., 4.

[193] Ibid., 4.

[194] "GCI's Introductory Heart Coherence Technique." HeartMath Institute. Accessed July 26, 2018. https://www.heartmath.org/gci/introductory-heart-coherence-technique/.

Chapter 16

Pearl Prihoda's PPIT Theory

"Transparency is all about letting in and embracing new ideas, new technology and new approaches."

— Gina McCarthy

I believe that embracing new ideas, technology, and approaches to sasquatch research is paramount if we are seeking truth while avidly tramping through the vast, wild, and incredibly beautiful wildernesses flourishing throughout the United States. Serious sasquatch researchers pursue an elusive truth, oftentimes with little support, and, frequently, producing questionable results. Their rigorous attention to detail and unbounded dedication, however, speak well for the entire sasquatch community. Pearl J. Prihoda is one of these valiant and persistent souls who has contributed to overall sasquatch research by creating a salient theory worthy of strong mention.

Unraveling Prihoda's PPIT Theory

Prihoda offered a clear statement at the start of her book, *Man-Otang: Bigfoot Myth or Reality*, published in 2014, which set the stage for sharing her decades-long research. She wrote, "Within this book I am going to show you that the potential or possibility for an undiscovered human or hybrid-human species to exist even today is not as far-fetched as some may think."[195] I wholeheartedly agree with her bold assessment. In addition, I believe my research offers staunch support for at least parts of her theory.

In my opinion, her work is a major contribution to the field of sasquatch research, and I feel compelled to include comments on some of Prihoda's thoughts about the intriguing and multifaceted theory she put forth. A primary reason I was initially attracted to Prihoda's theory in 2003 is because she showed an unerring dedication to her research. Prihoda, however, is not only an enthusiastic sasquatch researcher, she is also a historian for *Bigfootology*, selflessly digging out and storing a vast accumulation of information from ancient texts.

For instance, according to Rhettman A. Mullis, Jr., founder of Bigfootology,[196] Prihoda "finds historic artifacts that support the existence of Bigfoot throughout the millennia."[197] Bigfootology, founded in 2010, researches "much of the lore around the world in old and ancient cultures" in an effort to associate ancient, even archaic, myths and legends regarding so-called hairy creatures with today's hairy, human-like apes known by many as sasquatch.[198]

Again, Prihoda, a private sasquatch researcher living in the state of Washington, has engaged in myriad hours of credible academic and field research for several decades. Prihoda posits what I believe is a trustworthy and captivating theory regarding "bigfoot" experiencing a form of so-called induced transparency. She named her theory "Photosynthetic Piezoelectric Induced Transparency," or PPIT, for short. Prihoda's theory is enthralling and, I believe, academically sound on many levels. I personally consider it conceivable based on the evidence she presents intermixed with my private research results and experiences.

Okay. Let's jump right into the swirling waters of PPIT and see where the currents take us. Don't be surprised, though, if the swirling currents of PPIT return us eventually to the concept of extremely low frequencies (ELF), which I introduced in Chapter 2 and have discussed throughout the book. Let's begin with a brief look at the way Prihoda defines the primary premise of her hypothesis. Prihoda offered the following succinct introduction to her theory:

> Photosynthetic Piezoelectric Induced Transparency [PPIT] is a hypothesis concerning a natural avenue towards gaining biological transparency. With the consumption of critically interchangeable elements, *electromagnetic energies from the earth*, and *very low or extremely low frequencies* that mirror a given environment, it is proposed that transparency may be gained (author's emphasis).[199]

I willingly state—unconditionally, I might add—that my research with magnetism, electromagnetism, biomagnetism, geomagnetism, and ELF support her premise about electromagnetic energies and ELF. In fact, I'll gladly leap onto yet another thin, vibrantly swaying limb and say that I and another have *seen* the so-called induced transparency effect unfold before our very eyes. To wit, Brush-tracker and I had an incident of that nature occur in 1999 as discussed in Chapter 1, in 2004 as discussed in Chapter 3, and again in 2009, also discussed in Chapter 3. Additionally, I mentioned yet another extraordinary incident in Chapter 13 when I saw this phenomenon occur **at my base campsite when a female sasquatch walked away from me**!

I have not engaged in a thorough investigation of Prihoda's theory from a *chemical* perspective; I'll leave that to folks with a stronger background in chemistry than mine. I am, however, as you know by now, investigating and commenting on the ideas of *geomagnetism*, *electromagnetism*, and *ELF* playing vital and long-term roles in sasquatch behavior, unusual happenings, and other types of sasquatch interactions, including a likely form of induced transparency.

Also, as a reminder, I have evidence supporting my theory that ELF factors into sasquatch behavior. As such, I believe my research and experiences with sasquatches and ELF serve as a qualifier for me to comment on Prihoda's theory from a semi-intelligent perspective. First, though, I need to build a foundation for my comments.

To begin, electromagnetism is commonly defined as, "An interaction between electricity and magnetism, as when an electric current or a changing electric field generates a magnetic field, or when a changing magnetic field generates an electric field."[200] In other words, when current flows through a conductor—regardless of the type—a corresponding magnetic field is created. Conversely, when *movement* occurs *within* a magnetic field—such as a sound wave—a corresponding electrical current is created. That is, *changes* in either an electrical conductor or within a magnetic field *creates* the other. This idea is relatively simple; it is also key to this discussion as we move through this chapter, so please do keep it in mind.

I also have great interest in *electromagnetic waves* (light waves) lying both within the visible light spectrum and just outside it in the electromagnetic spectrum known as IR. Too, I have an interest in *sound waves*, specifically those which are known—not surprisingly—as *infrasound*. To be clear, infrasound waves lie *below* the typical range of human hearing.

I introduced geomagnetism in Chapter 14. This chapter presents information about electromagnetic induced transparency (EIT), waves and quantum interference, IR, infrasound, piezoelectricity, electron spin resonance, radiation, and PPIT. Applied to sasquatch research, each of these concepts introduce a potential for unusual happenings to occur with sasquatches. This leads to another proposal:

Proposal 20:

> Electromagnetic induced transparency, waves and quantum interference, infrared frequencies, infrasound, piezoelectricity, electron spin resonance, radiation, and PPIT work together, allowing sasquatches to engage in natural behaviors such as full induced or partial induced transparency.

I won't dispute that paranormal activity occurs—for indeed it might. Even so, I strongly suspect that some of the so-called paranormal activity associated with sasquatches stems from an interaction with natural phenomena in ways that give an appearance of the paranormal.

What is Electromagnetic Induced Transparency (EIT)?

In a very basic sense, EIT is "… a quantum interference effect that permits the propagation [dissemination] of light through an otherwise opaque atomic medium."[201] Or, if you will, EIT is a phenomenon which leads to something known as *destructive quantum interference*, and destructive quantum interference causes a seeming solid object to become transparent.[202] Obviously, the physics of EIT is a tad or so complex; however, we do not need to understand quantum complexities underlying EIT for us to grasp the basics.

For example, note that electromagnetic induced transparency requires destructive quantum interference for it to occur. Stated simply, "interference is a phenomenon in which two waves superpose to form a resultant wave of greater, lower, or the same amplitude."[203] Bottom line: When destructive quantum interference occurs, temporary transparency ensues. A change in known electromagnetic qualities triggers destructive quantum interference. Simple and intriguing, eh? Okay … maybe just intriguing.

Next, let's see if we can simplify even more what is meant by quantum interference in contrast to typical interference. To do that, we need to briefly explore the ideas of both waves and quantum interference to see how they work together to produce either constructive or destructive quantum inference. No worries; this will not take long.

Waves and Quantum Interference

Fundamentally, there are two classifications of waves that command our attention: mechanical waves and electromagnetic waves. The primary distinction between these two waves is simple: mechanical waves *must* have a medium through which to pass (e.g., air, water, steel, etc.), while electromagnetic waves are essentially free from this restriction.

Examples of mechanical waves include sound waves, water waves, seismic waves, and waves traveling through a spring, such as the well-known *Slinky* toy as it rolls happily down a staircase while youngsters squeal and clap with glee. Examples of electromagnetic waves include light waves, microwaves, radio waves, X-rays, and gamma rays.[204] Our interest here focuses primarily on electromagnetic waves, although X-rays and gamma rays also play a role in our theories about sasquatches. This is due to the way space weather, which includes X-rays and gamma rays, influences Earth's magnetosphere, hence, all living organisms or biological systems on Earth. Space weather also affects light waves, microwaves, and radio waves.

The terms frequency, wavelength, and energy describe electromagnetic energy. These three methods of description are mathematically intertwined, meaning if one of them is known, the other two are readily calculated. Although this book does not detail mathematical proofs related to waves and electromagnetic energy, it is important to recognize that such proofs exist for all claims made during this discussion on waves.

My intention here is sharing general ways frequency, wavelength, and energy interact to produce possible unusual occurrences with sasquatches. We should not need mathematical proofs to garner an understanding about over-all actions of these waves.

We do, however, need to define five key terms related to light waves: *reflection*, *absorption*, *diffraction*, *scatter*, and *refraction*. I rely on NASA's definitions and graphics to describe these terms.[205] A diagram appears immediately following each definition to give you an accompanying visual sense for it.

Reflection occurs when incoming light hits an object and then bounces off.[206]

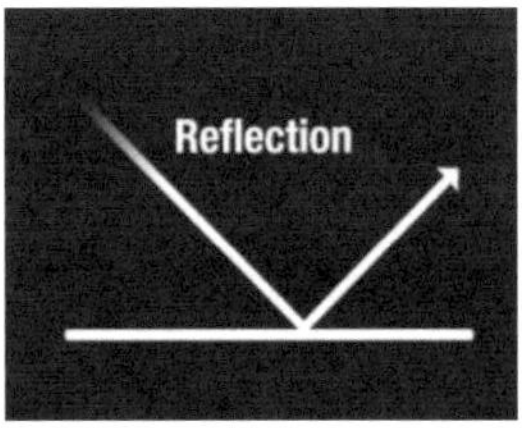

Diagram 16.1: Reflection — Credit: NASA

Absorption occurs when photons from incident light [light falling on an object] hit atoms and molecules and cause them to vibrate. The more an object's molecules move and vibrate, the hotter it becomes. This heat is then emitted from the object as thermal energy. [207] [**Note:** Heat and identification of a heat signature by sasquatches are important to this story.]

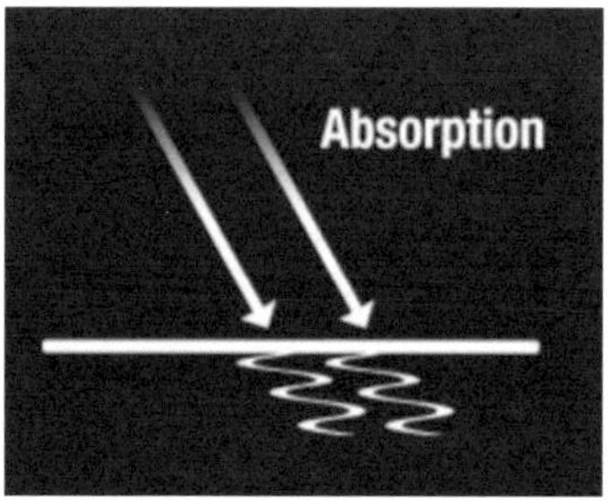

Diagram 16.2: Absorption — Credit: NASA

Diffraction is the bending and spreading of waves around an obstacle. [208]

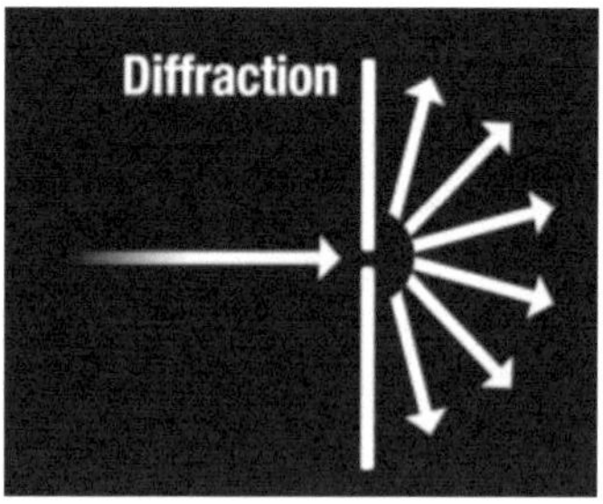

Diagram 16.3: Diffraction — Credit: NASA

Scattering occurs when light bounces off an object in a variety of directions. The amount of scattering that takes place depends on the wavelength of the light and the size and structure of the object. [209] [**Note:**

The concept of light scattering is important to Prihoda's PPIT hypothesis with respect to natural transparency.]

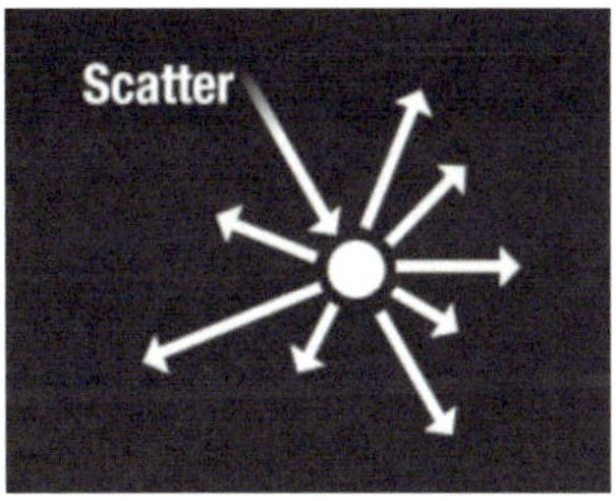

Diagram16.4: Scatter — Credit: NASA

Refraction is when light waves change direction as they pass from one medium to another. Light travels slower in air than in a vacuum, and even slower in water. As light travels into a different medium, the change in speed bends the light. Different wavelengths of light are slowed at different rates, which causes them to bend at different angles. [210] [**Note:** the concept of refraction also plays a key role in Prihoda's PPIT hypothesis with respect to natural transparency.]

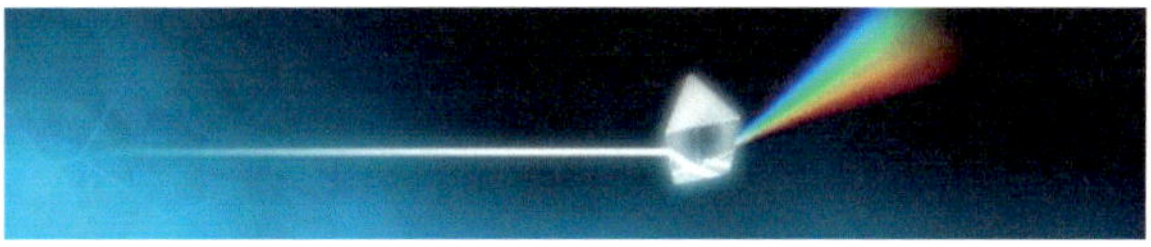

Diagram 16.5: Refraction — Credit: NASA

Now that we have definitions for basic actions occurring with light and light waves, let's drill down a bit into the idea of interference to see how quantum interference plays a role in potential natural transparency. As light waves interact, at least one of two types of quantum interference occurs: either **constructive** interference or **destructive** interference.

When two light waves interact, thereby creating a quantum interference, the interference is defined by something known as the superposition principle. For instance, the superposition principle simply informs us that when two waves interact, the resulting wave function (caused by the interference) becomes the sum of the two independent wave functions.[211] That is, when two waves interact, a type of interference ensues and the type of interference, constructive or destructive, depends on the resulting sum of the two waves.

I use diagram 16.6 to show a visual form of how interacting waves sum to produce a result. The result, of course, determines whether interacting waves produce constructive or destructive interference. For destructive interference to occur, the interacting waves need to be 180 degrees out-of-phase. Our interest here is on destructive interference while discussing EIT. Diagram 16.6 aids in visualization for a comparison between *in-phase* waves versus *out-of-phase* waves.

Let's begin, though, by discussing in-phase waves. Two in-phase waves of the same frequency can be in the same place at the same time.[212] When this occurs, the in-phase waves sum, producing a larger amplitude wave as seen at the top of diagram 16.6 A. You can readily recognize the in-phase wave relationship by looking at the peaks of each of the two smaller waves. Note how the peaks of each of the smaller waves line up.

Two in-phase waves appearing in the same place at the same time always add together to create a single larger amplitude wave as seen at the top of diagram 16.6 A. Therefore, constructive interference occurs when two in-phase waves add together to create a single larger amplitude wave.

Now look at diagram 16.6 B. Note how the peaks of the two smaller waves do **not** line up as do the ones in diagram 16.6 A. Rather, the peak of the bottom wave in diagram 16.6 B touches the valley of the upper wave. This is known as waves being out-of-phase; in this instance, 180 degrees out-of-phase.

This paragraph is important to our story because it describes the destructive interference that interests us. When two waves appear in the same place at the same time, and they are 180 degrees out-of-phase, based on the superposition principle, they still sum together. Because they are 180 degrees out-of-phase, though, one subtracts from the other, causing the sum to become zero. This is represented by the flat straight line at the top of diagram 16.6 B. Thus, *destructive interference* occurs when the sum of two *out-of-phase* waves equals zero.

A: Constructive Interference B: Destructive Interference

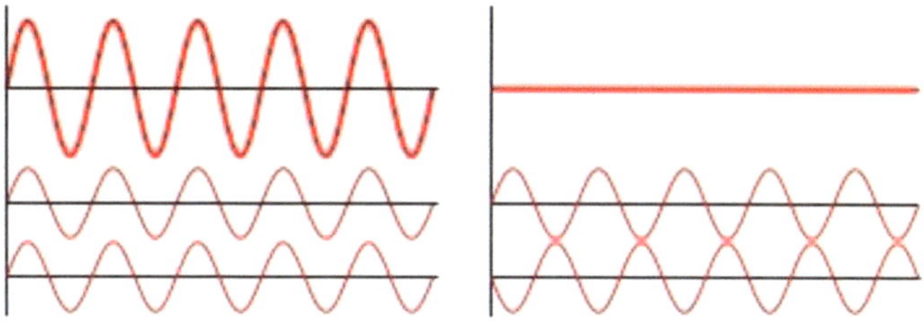

Diagram 16.6: Wave Interference[213]

Next, we need to peek a bit deeper into electromagnetic waves and frequencies. For example, the "Physics Classroom, Physics Tutorial" states, "Electromagnetic waves are produced by a vibrating electric charge and, as such, they consist of both an electric and a magnetic component."[214]

Prihoda's PPIT hypothesis relies in part on the way electromagnetic waves interact with matter. In that instance, she addresses very low frequency (VLF) and ELF waves. For example, she puts forth, "starting with low frequency radio waves, the human body is quite transparent [to the waves]."[215] Hmm! An interesting realization, eh?

In short, electromagnetic energies exist along a spectrum that holds a vast range of frequencies. For purposes of frequency identification, the *range* of frequencies are often displayed as specific regions within a total electromagnetic band. The electromagnetic spectrum contains the following bands of frequency:

1. **Radio:** AM Radio, Amateur Radio, Aircraft Communications
2. **Microwave:** Microwave Oven
3. **Infrared:** Remote Controls, Night Vision Goggles and Cameras
4. **Visible:** All things seen by the eye
5. **Ultraviolet:** Energy Emitted by our Sun
6. **X-Ray:** Medical, Airport Security, Universe Emitting Hot Gases

7. **Gamma Ray:** High-frequency electromagnetic radiation explosions.

The electromagnetic spectrum is presented visually in diagram 16.7.[216] Please note that the so-called visible spectrum is, in fact, a very narrow band within the total electromagnetic spectrum. This band appears as a vertical rainbow touching a cloud of light in diagram 16.7.

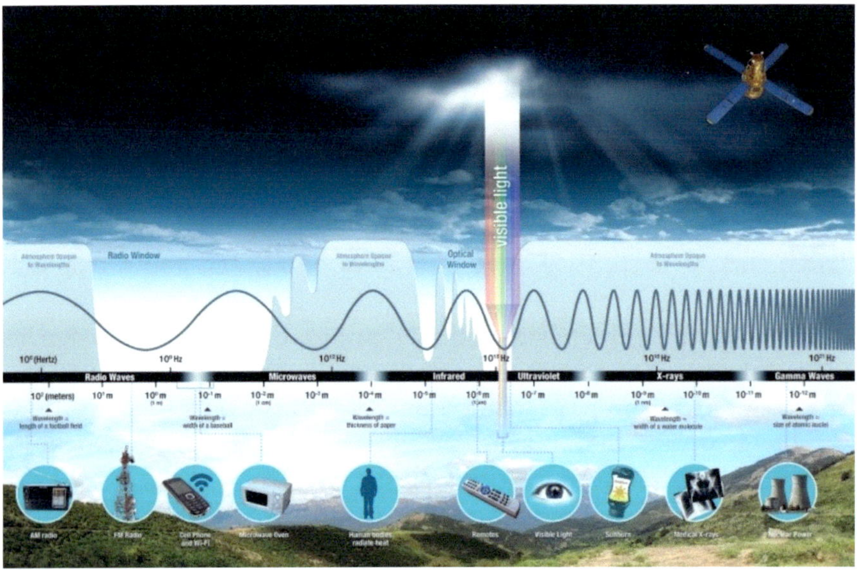

Diagram 16.7: The Electromagnetic Spectrum – Credit: NASA

The visible light spectrum exists between ultraviolet (UV) and IR frequencies. The visible light spectrum band depicts wavelengths visible to, or perceived by, the human eye. Diagram 16.8[217] shows a version of the visible light spectrum independent from the whole.

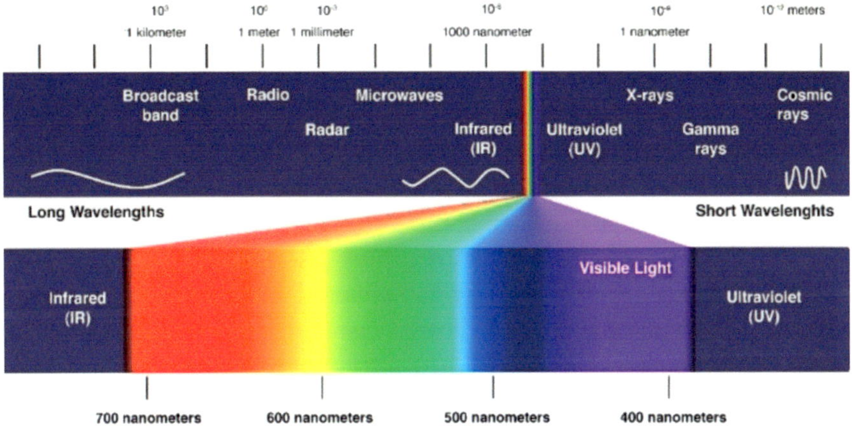

Diagram 16.8: Visible Light Spectrum — Credit NASA

Each of the colors within the visible light spectrum associate with light waves of specific lengths and frequencies. A common attribution with these light waves relates to how the *waves* of the colors grow *longer* as they move toward the left of the chart from UV to IR; conversely, the *frequency* of each color increases in speed and shortens in length as we move toward the right of diagrams 16.7-16.8. This means that from the left of Diagrams 16.7-16.8, we go from longer wavelengths, lower frequency and energy to shorter wavelengths, higher frequencies, and higher energies as we progress toward the right.

Next up, let's become a bit more familiar with the part of the electromagnetic spectrum known as *infrared*, or IR, which lies just below the red in the visible light spectrum. Infrared frequencies relate to this story of sasquatch and me in a couple of ways, which I'll get to shortly.

Infrared

Defined simply, IR frequencies within the electromagnetic spectrum fall below the red frequency, and are not visible to the human eye, although they do radiate heat, thereby producing a heat signature. Interestingly, infrared radiation is the segment of the electromagnetic spectrum we experience most in everyday life. As a rule, however, we are unaware of it because its waves are invisible to the human eye.

A notable factor associated with IR frequencies is the idea that people can feel infrared as heat.[218] Hmm! I believe sasquatches detect IR due to the heat signature it creates and emanates; furthermore, IR plays a role in Prihoda's PPIT theory.

To continue, most folks are familiar with TV remote controls. Every time you use a remote control, you employ principles of IR technology; trail cameras and other nighttime cameras and various motion sensors also use IR technology. Most dedicated sasquatch researchers are starkly aware that sasquatches are well-known to avoid the darn camera traps. Trail cams using IR technology emit a red glow as well as heat; both the red glow and heat signature seem readily detectable by sasquatches. Bummer!

Some of the newer trail cams purport to be no glow. I used one of these during my 2017 research. Even without a glow, bipedal anomalies detected and either avoided or interfered with the trail cam. Again, I suspect the issue of infrared technology emitting heat has a bearing on how sasquatches detect them.

Another stimulating use for IR is found in Infrared Reflectography (IRR). IRR was developed in the late 1960s by Dutch physicist J. R. J. van Asperen De Boer as a technique to "look through paint layers."[219] Infrared Reflectography is pertinent to our discussion because "when the longer wavelengths of infrared radiation penetrate paint layers, the upper layers appear transparent."[220] In fact, "the longer the wavelength of the infrared … the easier it is to penetrate to the layers beneath."[221] Again, long wavelengths, low frequency, and low energy are used to induce a form of transparency, this time with upper layers of paint on a painting becoming transparent, thereby allowing under layers to be clearly viewed.

Although I'm sure this aids in the determination of fraudulent paintings or in the discovery of paintings under others, I also believe IRR offers yet another reference to the ideas of IR frequencies and induced transparency. Granted, IR reflectography is not overtly relevant to this overall discussion; however, it certainly seems tangentially interesting and applicable to me. The central theme here relates to how longer wavelengths, lower frequency, and lower energy assists induced transparency.

Next up we peek at the idea of infrasound to see if that part of the electromagnetic spectrum serves a purpose in our research efforts. I, of course, believe it does. No surprise, there, eh?

Infrasound

A simple definition for infrasound is available from the *Merriam-Webster Medical Dictionary*: "a wave phenomenon of the same physical nature as sound, but with frequencies below the range of human hearing."[222] Sound familiar? This is obviously quite similar to infrared, where IR frequencies fall below the range of human vision.

Recall from an earlier discussion that sound waves are mechanical waves and need a medium through which to pass. This aspect of infrasound differs from IR, because IR waves are not bound by this restriction. A clear distinction for sure, albeit not an issue or concern to this discussion.

Historic references of the human sound range places the lower end at 20 Hz.[223-224] More recent studies suggest that the lower range is 16 Hz.[225] I have no interest in quibbling about the 4 Hz known to exist between older and newer studies, although this might be important were we to delve *deeply* into the science of this subject. Rather, my interest lies in how frequencies below the low end—be it 16 or 20 Hz—might intentionally be used by sasquatches to affect humans, perhaps in a negative or uncomfortable way.

There are a variety of sources of infrasound. For example, diagram 16.9[226] shows a visual of several of these sources. Spend a moment with this diagram and you might find yourself surprised at the sources of infrasound. For instance, who knew that meteors produced infrasound? Or that microbaroms, a type of "atmospheric infrasonic wave generated in marine storms"[227] caused infrasonic sound? So much to learn!

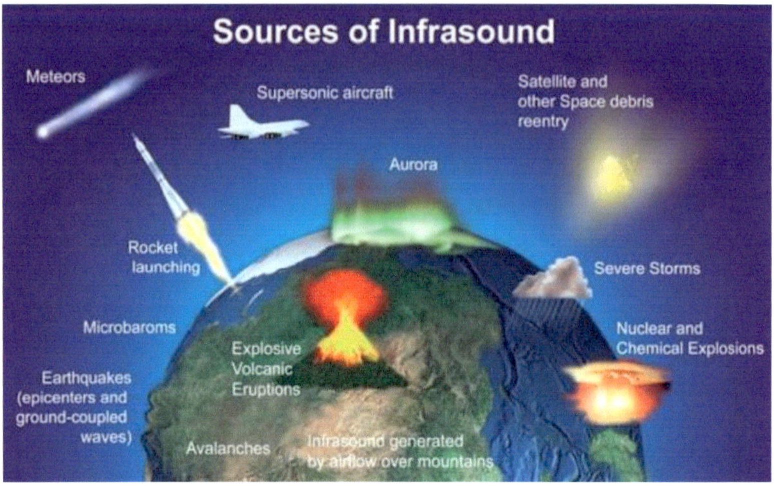

Diagram 16.9: Sources of Infrasound — Credit: NASA

In addition to the sources portrayed in diagram 16.9, it turns out many animals *emit* or *interact* with infrasound in known ways. For example, whales, elephants, rhinoceroses, hippopotamuses, okapi, giraffes, and dogs emit infrasound.[228] Other species that emit or otherwise interact with infrasound are crickets, humming birds in flight, katydids and grasshoppers, mosquitoes, and bats. Interestingly, researcher David Larom, et al, wrote, "Many animal species use long-range [infrasound] calls to *establish their use of space and their relationships with members of their own and other species* (author's emphasis)."[229]

Even though sasquatches are not mentioned among the mammals that emit and/or interact with infrasound, it is assuredly a large mammal. This, then, prompts another proposal:

Proposal 21:

> Sasquatches use infrasound as part of their communication repertoire, most likely for defense or to stay in touch with other members of a family, tribe or community.

Once again, each waveform of infrasound presents as a frequency, as do IR frequencies. The primary difference between the two relates to sound needing a medium through which to travel, while IR frequencies do not. Further, because sasquatches can detect infrared frequencies, it shouldn't take a huge leap of logic to accept that they also detect or make practical use of infrasound. Frankly, as mentioned above, I strongly suspect sasquatches *emit* infrasound as part of a communication for defense, or as a location mechanism.

Infrasound occurs as low frequency, and *low frequency noise* in general is worthy of an extra paragraph or two because of the way it affects living organisms. To wit, "Low frequency noise annoyance is related to headaches, unusual tiredness, lack of concentration, irritation, and pressure on the eardrum."[230] Furthermore, many studies suggest sleep may be adversely affected when experiencing low frequency noise; therefore, although we as humans are not able to detect low frequency noise, such noises can—and allegedly do—have certain adverse effects on the human population.

To emphasize the importance of being aware of some of the risks attributed to low frequency noise, let me remind you of a situation that occurred in Cuba beginning around the fall of 2016. The *Associated Press* wrote an article about an incident that occurred in Cuba; it stated:

> After months of investigation, U.S. officials concluded that … diplomats had been exposed to an advanced device that operated outside the range of audible sound and had been deployed either inside or outside their residences.[231]

Likewise, an article published by the *Independent* in the United Kingdom reported "permanent hearing loss was another diagnosis, and that additional symptoms had included brain swelling, severe headaches, loss of balance and cognitive disruption."[232] Hmm! Low frequency noise can apparently be naughty enough to cause cognitive disruption, which tangentially relates to cognitive dissonance during an anomalous experience. It just keeps getting more and more interesting, eh?

Remarkably, as I write this section, another incident of a diplomat suffering from low frequency noise has occurred; this time in China. *The*

Diplomat, the premier international current affairs magazine in the Asia-Pacific region, published an article on May 29, 2018, claiming a U.S. government employee "recently reported subtle and vague, but abnormal, sensations of sound and pressure."[233] Additionally, the *New York Times* also wrote a piece about American diplomats suffering from low frequency noise issues in China. Clearly, and regardless of the cause, adverse health experienced by American diplomats in both Cuba and China speak volumes about infrasound and its potential negative effects on human health.

This discussion returns me to the ideas of cognitive dissonance and cognitive disruption. While cognitive dissonance and cognitive disruption are not the same, one might argue that they share an oblique association. For instance, cognitive dissonance can cause a sense of cognitive disruption, which is known to cause confusion, delusions, hallucinations, depersonalization, impulsivity, and even an inability to comprehend and formulate language.[234] With that said, I offer another proposal:

Proposal 22:

> Sasquatches emit low frequency sounds around humans to intentionally create a sense of cognitive disruption or other types of dysfunction in humans, thereby effectively chasing them away from their territory.

Cognitive dissonance, on the other hand, although occurring during a sasquatch encounter, is **not** caused by sasquatches; rather, it is due to the way one reacts to an unexpected and highly startling anomalous experience. This is an important distinction. Also, note how a sense of cognitive disruption in Cuba and China diplomats caused them to vacate Cuba's and China's territories. Hmm!

Awareness that low frequency noise can be harmful to living organisms under certain conditions dovetails rather nicely into this story of sasquatch and me. For instance, in keeping with proposal 22, many reports exist where witnesses claim they experienced headaches, dizziness, and temporary loss of mental clarity, or if you will, minor cognitive disruption at the time of sasquatch encounters. Such effects could well have been caused when sasquatches emitted low frequency noises! Noises which fall below typical human hearing ranges would not be noticed by those having the negative experience. Regardless, I believe low frequency noise is a decisive factor well worth consideration while researching sasquatches. Moreover, the type of experience mentioned above falls under the umbrella of an anomalous experience causing cognitive dissonance. I implement coherence techniques to avoid such unpleasant experiences—and it works well!

Continuing with sound, Prihoda suggested that sound and harmonics, VLF and ELF may be attributed to sasquatches, and she presented the idea of so-called *throat singers* as examples.[235] For instance, she mentioned the Throat Singers of Tuva (Tuva is found near the region of the Siberian/Mongolian border) and Inuit Throat Singing (found from Northern Alaska to Greenland). She reminded us that throat singers can sing more than one note at the same time, Furthermore, she claims that "some [throat] singers can sing up to six different notes at once!"[236] Intriguing, to say the least!

You can listen online to *The Throat Singers of Tuva* by checking the next two endnotes and following their URLs.[237-238] In addition, you may hear *Inuit Throat Singing* online by checking the next two endnotes and following their URLs.[239-240]

I believe I've heard sasquatches make low guttural sounds akin to those made by throat singers. I do not, however, recall hearing any of the highest-pitched sounds used by throat singers being uttered by sasquatches. Even so, I believe sasquatches do have throat singing ability—or something comparable—that they use for communication or, perhaps, at community dances? You know, the Saturday night dance thing on the banks of a local river mentioned in Chapter 5? Anyway, next on deck is a discussion on piezoelectricity.

Piezoelectricity

The name Prihoda uses for her PPIT theory includes the word *piezoelectricity*. Hence, piezoelectricity needs to be addressed prior to delving into a summary of her theory.

Piezoelectricity makes use of *pressure* to cause a change in a system. According to Wikipedia, "Piezoelectricity is the electric charge that accumulates in certain solid materials (such as crystals, certain ceramics, and biological matter such as bone, DNA and various proteins)."[241] Already this stuff should be catching your attention. Why? First, you may recall we talked about crystals, electromagnetic energies, and charge (i.e., capacitance) in Chapters 6 and 7. Second, learning that piezoelectricity relates to biological matter, including bones, DNA, and various proteins should cause your heart to pump with anticipation about what's coming up next!

Sharon Whale, known for her use of electronic gem therapy, wrote, "Piezo electricity [sic] relates to … charges in quartz crystal when subjected to mechanical pressure (squeezing and tapping etc.). The crystal will transduce [convert] the mechanical frequencies of the stress into electrical frequencies and voltages."[242] As a point of interest, the electrical frequencies and voltages Whale speaks of can be empirically measured with appropriate equipment.

The important points to take from this part of the discussion are first, a piezoelectric effect occurs with crystals and biological matter as well as various proteins, and second, either squeezing or tapping a quartz crystal will create an electromagnetic effect. The latter feature of quartz crystals relates to this story of sasquatch and me. For instance, recall from Chapter 6 our discussion on quartz crystals and a possible relationship between them and sasquatch activities along Dimension Lane. We also mentioned quartz crystals on other occasions, specifically in Chapters 7 and 9.

We know that sasquatches engage in an activity termed wood knocking. I suspect that sasquatches use variations of this—which creates vibrations that affect crystals—to activate electromagnetic waves from the quartz crystals for a particular purpose. If a form of tapping is used, remember that sound is created, and sound requires a medium through which to pass. This would, of course, include infrasound. The medium in the instance of Dimension Lane may very well have been crystals.

HeartMath and other researchers have adequately proven electromagnetic energies interacting with chemical changes affect both the physical heart and brain energies. Some of the changes that occur could cause physical discomfort and cognitive disruption, or, perhaps other types of disorientation to humans and animals. This would be the opposite effect of coherence, which, of course, we would rather not experience.

We have shown that the human heart and brain emit measurable quantities of electromagnetic frequencies detected by others around us. Furthermore, we also know that when humans and animals are near each other, exchanges of these energies occur.

With this information under our belts, I propose the following:

Proposal 23:

> Sasquatches use tapping, wood banging, rock banging, or other methods to create sound waves to alter natural electromagnetic qualities of quartz crystals, creating ELF electromagnetic energies to use for their specific benefit, quite possibly to assist with induced transparency.

Proposal 24:

> When hydrogen sulfide (H2S) is intentionally manipulated by sasquatches—as suggested by Prihoda—they can emanate an unseen odoriferous miasma to disorient humans or animals, which serves as a nasty yet effective defense mechanism.

Potassium and Rubidium

Next, it's easy to see why Prihoda believes both potassium and rubidium play key roles in her PPIT theory. For example, the United

Kingdom's Royal Society of Chemistry wrote, "Potassium is essential to life. Potassium ions are found in all cells. It is important for maintaining fluid and electrolyte balance."[243] Rubidium, on the other hand, "… has no known biological role and is non-toxic," and "because of its chemical similarity to potassium we absorb it from our food, and the average person has stores of about half a gram."[244]

You may recall our discussion from Chapter 7 about missing dirt, and how analysis of the sampled dirt showed potassium, magnesium, and calcium as part of it. I believe sasquatches were eating that dirt to supplement their intake of the alkali metal potassium, and quite possibly ingesting magnesium and calcium supplements as well. Clearly, humans ingest myriad quantities of supplements each year; one estimate exceeded 20 billion dollars' worth for 2015. Heck, sasquatches seem to get their supplements for free! Smart! Is it time to engage freely in geophagia?

Another interesting feature about both potassium and rubidium is that they are easily ionized because each has only one electron in its outer ring. When ionizing radiation plays *whack-an-atom* with potassium or rubidium atoms—as discussed in Chapter 15—the single electron in their outer rings is easily dislodged.

Once the outer ring electron is dislodged, the charge of the atom changes, creating a positive ion called a cation. Prihoda claims this anomaly is the reason both potassium and rubidium play crucial roles in PPIT. Diagrams 16.10 and 16.11 show a single electron (red) in the outer ring of potassium and rubidium atoms. The orange circle in the center of each atom stands for the nucleus. The outer ring electron is easily influenced because it exists alone in the outer ring. Again, photos and diagrams appearing in this book will soon be available in color on my website sasquatchandme.com.

Okay. We've figured out that ionization strips off the single electron in the outer ring of a potassium or rubidium atom, causing them to become positive ions, or cations. Let's see if we can figure out why Prihoda considers this action important to her PPIT hypothesis.

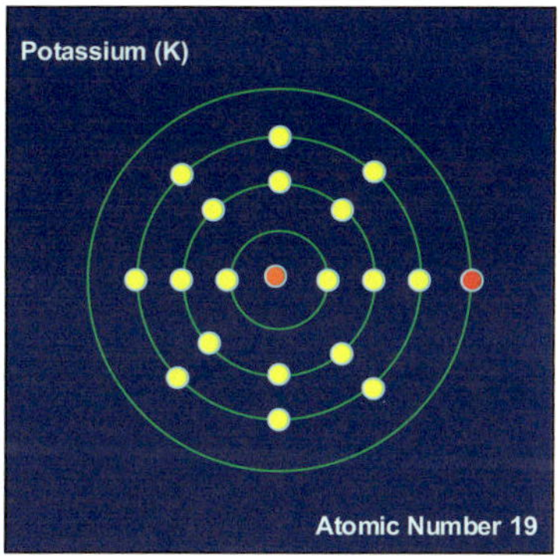

Diagram 16.10: Basic Structure of Potassium Atom

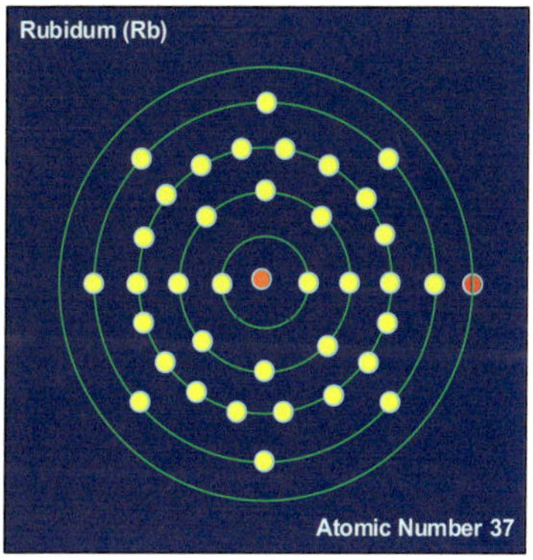

Diagram 16.11: Basic Structure of Rubidium Atom

To understand why Prihoda considers this a key factor to her theory, we must also examine premises related to an electron magnetic dipole moment, as well as something known as electron spin resonance (ESR). Yuk! Actually, this discussion will be quick and relatively painless.

Magnetic Dipole Moment and Electron Spin Resonance (ESR)

We need simple definitions for each of these phrases. First, though, let's define terms associated with ESR.

Electron: A tiny particle of matter smaller than an atom with a negative electrical charge.[245] [The fact that electrons hold a negative electric charge is important to remember.]
Magnetic Dipole Moment: A magnetic dipole moment is often simply called a magnetic moment; both refer to the same action. In the end, when the "onion skin" of a magnetic dipole moment is peeled away, it stands for a quantity of the magnetic strength and the orientation of a magnet that produces a magnetic field.[246] [I believe that the magnetic strength along with orientation of the magnetic lines of force controlling the strength are relevant factors in Prihoda's theory.]
Electron Spin Resonance (ESR): Wikipedia informs us that, due to spin, "unpaired electrons have a magnetic dipole moment and act like tiny magnets."[247] [The single electron in the outer rings of potassium and rubidium qualify as unpaired electrons.] We can view ESR as a technique to investigate paramagnetic substances by subjecting them to high-frequency radiation in a strong magnetic field. Finally, spin changes of single electrons cause radiation to be absorbed at certain frequencies.[248] [Absorption of radiation is another key factor in Prihoda's PPIT theory.]
Paramagnetism occurs in certain materials and elements, such as potassium and rubidium, which are both alkali metals. Paramagnetic materials are weakly attracted by an external magnetic field. In short, paramagnetism occurs due to unpaired electron behavior. Also, paramagnetic materials cannot be permanently magnetized; rather, whatever magnetism they experience is always temporary. This factor plays a leading role in ESR.

Diagram 16.12 shows a rough facsimile of electron spin. It turns out that electron spin follows the magnetic field lines of a dipole magnet. Interesting how we seem to always return to the idea of a dipole magnet during our discussions about sasquatch research, eh?

At any rate, an electron spins from the *north polarity* to the *south polarity* and only moves in two directions: *up* or *down*. When an electron moves from *down-to-up* (as shown in the left side of diagram 16.12), electron spin is *counterclockwise*. When it moves from *up-to-down* (as shown in the right side of diagram 16.12), electron spin is *clockwise*. The direction of spin plays a key role in Prihoda's theory. Too, direction of electron spin is central to my research, particularly with respect to sasquatch behavior and specific unusual activities. You will see what I mean by this shortly.

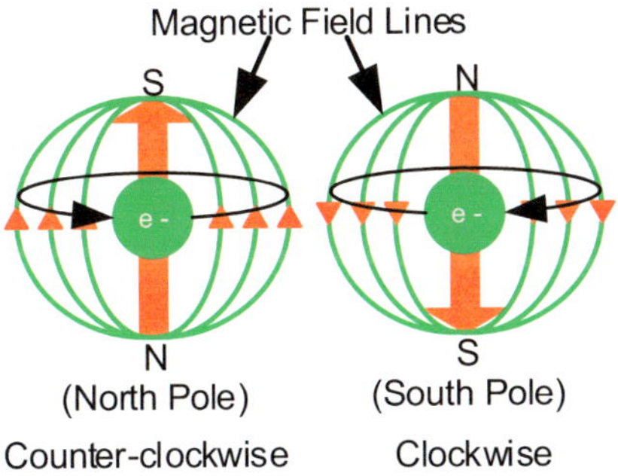

Diagram 16.12: Electron Spin — Credit PYFN

Prihoda informed us, "… the North Pole counter-clockwise spin [down-to-up] provides a negative form of energy and slows down cellular activity, while just the opposite will occur with a clockwise [up-to-down] spin."[249] You might wonder why this is important. Prihoda wrote, "It only takes a very low intensity magnetic field to effect chemical reactions [for example, in potassium and rubidium] and these reactions have a definite biological effect on your body."[250] In other words, Prihoda is suggesting that a combination of chemical reactions intermingled with low intensity magnetic fields (i.e., ELFs) while undergoing ESR contribute to temporary induced transparency. Wow!

In a nutshell, electron spin creates a magnetic dipole moment, magnetizing an event. The strength of the magnetic dipole moment contributes to the result. In this instance, the result is a form of induced transparency. How cool is this?

Elaboration on behavioral anomalies discussed in Chapter 4

You may recall from Chapter 4 that we had visitors our first night in camp. Several photos showed somewhat circular grass mattings where something had been. I believe these photos speak to the issue of electron spin resonance, and photo 4.4 is reproduced below as photo 16.1. My focus here relates to the *direction* of the matted grass, for I believe the direction of the grass shows a form of unusual sasquatch activity.

Photo 16.1: A circular matted grass area

Looking closely, you might notice from photo 16.1 that the *underside* of the grass appears as *clockwise*, while the *upper* part of it looks to be in a *counterclockwise* position. The clockwise and counterclockwise indicators fit directional and activity patterns related to electron spin. That is, when electron spin moves from *up-to-down*, the direction of spin is *clockwise*. On the other hand, when it moves from *down-to-up*, the spin is *counterclockwise*. Photo 16.1 shows both effects; granted, you must look closely to see it.

Now for the weird part. Assuming sasquatches interact with electromagnetic light waves influenced with chemical reactions, a question forms based on the clockwise/counterclockwise positioning of the matted grasses. First, I believe it is possible for sasquatches to use electromagnetic energies intermingled with chemical interactions during a form of induced transparency as they *enter* or *leave* a location *unseen*. That is, a *clockwise* electron spin resonance might bring an induced transparent sasquatch *into* a location while a *counterclockwise* electron spin resonance might take an induced transparent sasquatch *out* or away from a location. I believe that neither of these activities would be visual to humans. This, of course, leads to another proposal:

Proposal 25:

> Sasquatches use a combination of electromagnetic light waves, chemical interactions, and electron spin physics to move into and/or out of a specific location unseen; this is another form of induced transparency.

If sasquatches can engage in this type of ingress/egress, the matted grasses seen in photo 16.1 show such actions. Additionally, I have photos from many years of research showing the same or similar type of activity as seen in photo 16.1. Once again, merely additional food for thought.

Radiation Can Change DNA Molecules

Recall from earlier discussions the idea that ionizing radiation plays *whack-an-atom* with unpaired atoms. Evidence exists that radiation can—and does at times—destroy or otherwise adversely affect DNA molecules. For instance, NASA wrote, "Radiation can sever one or both strands of the molecule, forming single or double strand breaks."[251] Diagram 16.13 shows a normal double-helix DNA molecule on the left and one damaged from radiation on the right.

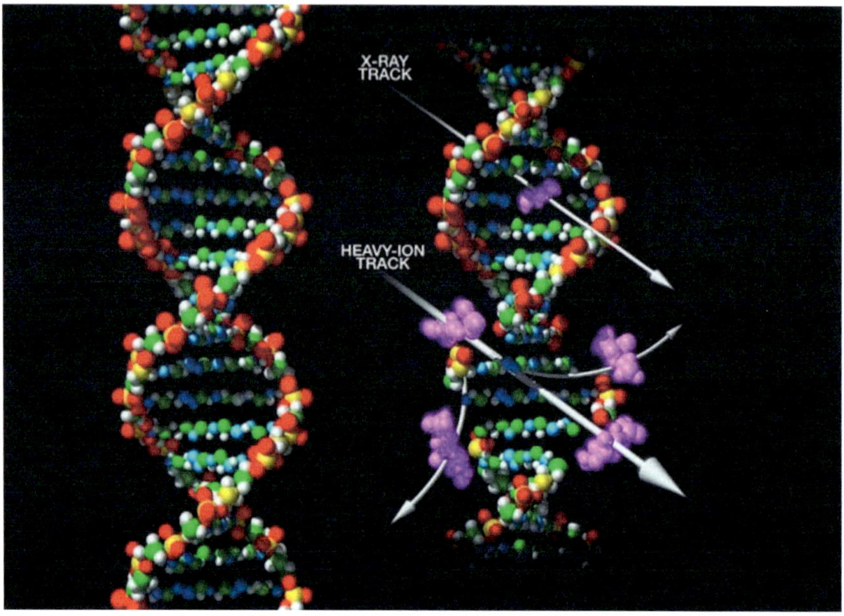

Diagram 16.13: Lft: Normal Rgt: Damaged — Credit: NASA

The point of including this information is simple. I want you to visually relate an impression of radiation to the concept of molecular level alterations due to radiation. The ideas of ELF, magnetism, geomagnetism, space weather, X-rays, gamma rays, cosmic rays, PPIT, and so forth, each and all deal with radiation of one form or another. In the end, it is well known that radiation affects cellular level activity, including DNA.

It is also important to realize that radiative alterations to DNA are not always harmful. In fact, positive changes to molecular structures can be as frequent as damaging ones. Therefore, it is entirely conceivable for positive or otherwise helpful radiative alterations to DNA to coincide with one form or another of induced transparency.

This isn't the only astonishing thing about DNA. Lo and behold, an online news article, written by Kristen V. Brown, senior writer for *Gizmodo*, was published on April 23, 2018. It touted another shocking claim made as a result of a then new study: "Forget the Double Helix—Scientists Discovered a New DNA Structure Inside Human Cells."[252] The study, titled "I-motif structures are formed in the nuclei of human cells," offers startling information about this new discovery. For example, it concludes that the new DNA structure, called an *i-motif structure*, forms in the nuclei of human cells. Dang! The nuclei of human cells are seriously itty-bitty tiny. Gotta love microscopic viewing instruments! Finally, the new DNA structure appears almost as a twisted knot rather than the familiar double helix.

If nothing else, the i-motif study should open our eyes and minds to the idea that all living organisms are profoundly and intricately complex. In some ways, we as humans are not only unknown to ourselves, we are also capable of delightfully surprising ourselves with ever forward-reaching knowledge about the whole of all that we are, and the yet to be discovered ways we interact with all that is around us.

I believe the same is true for sasquatch research. That is, so much about sasquatches is currently unknown to us; yet, it seems that dedicated research by many private investigators opens new doors for us when sharing their long-term data. Such is true for Prihoda and, I hope, for this book as well.

Continuing, I offer another thought-provoking note about DNA. According to *Express*, an online British news outlet, NASA is preparing a so-called manned mission to Mars. One of its major concerns is how to protect an astronaut's DNA from severe radiation damage during the trip. Amazingly, NASA is clinically striving to discover ways to *intentionally* alter human DNA in astronauts *before* the trip. NASA scientists believe that changing an astronaut's DNA is a conceivable way to avoid damaging cancer radiation during the trip.[253] Really! Check it out. And the possibility of temporary induced transparency is bizarre? Goodness!

Staying abreast of current science can sometimes befuddle the mind and stymie our thought processes due to shocking new information. Researching sasquatches and possible induced transparency, on the other hand, is invigorating and perhaps, far less bizarre than some of the more recent science or high-tech and artificial intelligence (AI) discoveries! Maybe NASA should peek at the bountiful and profound sasquatch research floating around, eh? If even a little of this stuff is valid, I'm confident NASA could make decent use of it! Smile!

Finally, I strongly suspect that a researcher with enough funding will one day appear from the ethereal mists of the horizon, stride blithely into this forum and then confirm and validate the induced transparency theory. Although the converse is also a possibility, I really believe that most of the underlying premises upon which Prihoda's hypothesis is formulated and what my personal research suggests can be confirmed with proper funding, organization, and due diligence in research.

Summary of PPIT Important Information

Recall from earlier discussions that every cell within the human body has one form or another of electromagnetic energy. Further, each cell within carbon-based creatures, including humans and sasquatches, continuously interacts with chemical properties of other cells, as well as being susceptible to external influences. These known variables keep the medical community perpetually busy sorting through actions/reactions as newer and higher technology continually produces greater insights into the world of human health. Amazingly, it is not beyond the realm of possibilities that future health care will include instruments commonly seen in science fiction TV series or movies. Good or bad? Right or wrong? I lack the wisdom to address these questions.

Prihoda claims chemical exchanges within (and, at times, without) the body interact with electromagnetism—VLF and ELF frequencies. She believes these interactions combine in ways to create a possible Photosynthetic Piezoelectric Induced Transparency.

Additionally, Prihoda did a reputable job discussing how the concepts of zeolites, electron spin, and polymers affect induced transparency. She informed us that DNA (Deoxyribonucleic acid) is a polymer, as are our bones and muscle fibers. She also wrote, "Virtually all organisms give off resonance & light."[254] She similarly posited, "Potassium/Rubidium and light all act as active neural transmitters…"[255]

This is important information because neurotransmitters send chemical messages between neurons. Neurotransmitters may be defined as "… the molecules used by the nervous system to transmit messages between neurons, or from neurons to muscles."[256] Potassium, rubidium, and light, also play roles in Prihoda's theory; in essence, they are energetic neural transmitters!

Whew! Sure, and I hope some of this stuff is beginning to come together for you. I've pursued these lines of thought for years and still don't have a complete understanding of them. Even so, I hope that this story aids a bit in showing how I've arrived at the proposals and suppositions presented within this book.

Continuing, Prihoda further postulated, "Attaining transparent skin has to do with the fact that light is scattered as it moves through the skin."[257] She also claims, "PPIT has to do with the *way* in which the light is scattered, [by] *matching the refractive index of the given environment* (author's emphasis)."[258] Fascinating and very telling!

We earlier defined both scattering and refraction. Recall that *scattering* occurs when light bounces off an object in many directions, while *refraction* ensues when light waves change direction as they pass from one medium to another. If Prihoda is correct—and she might well be—for transparency to occur in tissue, "light must not be absorbed nor reflected;"[259] rather, light waves should *scatter* (bounce off an object in a variety of directions) and *refract* (change direction as they pass from one medium to another). It turns out that light waves can both scatter and refract quite easily, and if they interact simultaneously with certain chemicals—induced transparency occurs! Terribly transparent, wouldn't you say? Smile!

Okay, while this chapter sadly lacks in granting wide-ranging justice to Prihoda's PPIT theory, I believe it offers enough insight into her hypothesis to understand why I support it. Be still my beating heart! This means we are nearing an end to this part of our journey about sasquatch and me.

Next up is chapter 17, where I share semi-conclusive thoughts; semi-conclusive because this is sasquatch research and absolute definitive conclusions have not yet come into clear view. Rather, they seem to be hanging out on that ever-ethereal horizon, shimmering brightly yet indistinct while teasing and tantalizing us to move ever closer to clear and sound realizations about the mysterious sasquatches we find so beguiling and intriguing.

Finally, Chapter 18 presents a fleeting peek into future research activities. Chapter 19 closes this story by offering many of my personal beliefs about sasquatches.

Chapter 16

[195] Prihoda, Pearl J. *Man-Otang: Bigfoot Myth or Reality*. Bellevue: SAFEHouse Publishing, WA, 2014, 17.
[196] "Bigfootology" is a trademarked word.
[197] Ibid., vii.
[198] Ibid., vii-viii.
[199] Ibid., 203.
[200] Staff. "Electromagnetism." One-dimensional Dictionary Definition. Accessed May 23, 2018. http://www.yourdictionary.com/electromagnetism.
[201] Liu, Chien, Zachary Dutton, Cyrus H. Behroozi, and Lene Vastergaard Hau. "Observation of Coherent Optical Information Storage in Anatomic Medium Using Halted Light Pulses." ReadCube for Researchers. January 25, 2001. Accessed April 27, 2018. doi:10.1038/35054017.
[202] Phillips, Mark Christopher. *Electromagnetically Induced Transparency in Semiconductors*. Master's thesis, University of Oregon, Department of Physics and the Graduate School, 2002. Accessed May 23, 2018, iii.
[203] "Wave Interference." Wikipedia. May 27, 2018. Accessed June 02, 2018. https://en.wikipedia.org/wiki/Wave_interference#Quantum_interference.
[204] Staff. "Physics for Kids." Ducksters Educational Site. Accessed May 01, 2018. http://www.ducksters.com/science/physics/waves.php.
[205] NASA. "Wave Behaviors." NASA: Science Beta. Accessed May 24, 2018. https://science.nasa.gov/ems/03_behaviors.
[206] Ibid.
[207] Ibid.
[208] Ibid.

[209] Ibid.
[210] Ibid.
[211] Jones, Andrew Zimmerman. "Interference, Diffraction & the Principle of Superposition." ThoughtCo. Accessed April 27, 2018 https://www.thoughtco.com/interference-diffraction-principle-of-superposition-2699048.
[212] "5. Waves and Interference." The Music of Physics. Accessed April 30, 2018. http://www.phys.uconn.edu/~gibson/Notes/Website.htm.
[213] "Wave Interference." Wikipedia. Wikimedia Foundation, March 4, 2019. https://en.wikipedia.org/wiki/Wave_interference.
[214] Staff. "The Electromagnetic and Visible Spectra." The Physics Classroom. Accessed May 02, 2018. http://www.physicsclassroom.com/class/light/Lesson-2/The-Electromagnetic-and-Visible-Spectra.
[215] Prihoda, Pearl J. *Man-Otang: Bigfoot Myth or Reality*. Bellevue: SAFEHouse Publishing, WA, 2014, 166.
[216] NASA, Staff. "Introduction to the Electromagnetic Spectrum." *NASA: Tour of the Electromagnetic Spectrum*, NASA, 2010. science.nasa.gov/ems/01_intro.
[217] "Pics about Space." *Space Shuttle Challenger Disaster Bodies - Pics about Space*, Pics about Space. https://pics-about-space.com/em-spectrum-nasa?p=1.
[218] Lucas, Jim. "What Is Infrared?" LiveScience. March 26, 2015. Accessed May 21, 2018. https://www.livescience.com/50260-infrared-radiation.html.
[219] Staff. "Infrared Reflectography." The Art Institute of Chicago. Accessed May 26, 2018. http://www.artic.edu/collections/conservation/revealing-picasso-conservation-project/examination-techniques/infrared.
[220] Ibid.
[221] Ibid.
[222] Staff. "Infrasound Medical Definition." Merriam-Webster. Accessed May 04, 2018. https://www.merriam-webster.com/medical/infrasound.
[223] Madankan, Zahra, Noushin Riahi, and Akbar Ranjbar. "Infrasound Source Identification Based on Spectral Moment Features." *10.11648.j.ijiis.20160503.11* 5, no. 3 (April 26, 2016): 37-41. Accessed May 4, 2018. doi:10.11648/j.ijiis.20160503.11.

[224] Victoria. "Infrasound Can Mess with Your Head." *NeuroNotes: Neuro Research Project* (blog), October 12, 2017. Accessed May 04, 2018. https://neuroresearchproject.com/2013/02/19/1289.

[225] Leventhall, Geoff. "What is Infrasound?" *Progress in Biophysics and Molecular Biology* 93:130-137. Accessed May 4, 2018. https://www.academia.edu/7333731/What_is_infrasound..

[226] University of Western Ontario. "Meteor Infrasound." Meteor Physics. December 14, 2011. Accessed May 04, 2018. http://meteor.uwo.ca/research/infrasound/is_whatisIS.html.

[227] "Microbarom." Your Dictionary. Accessed March 22, 2018. https://www.yourdictionary.com/microbarom.

[228] Staff. "What Animals Use Infrasound?" Reference.com. Accessed May 04, 2018. https://www.reference.com/science/animals-use-infrasound-a7b71a3517ef3966#.

[229] Larom, David, Michael Garstang, Katherine Payne, Richard Raspet, and Malan Lindeque. "The Influence of Surface Atmospheric Conditions on the Range and Area Reached by Animal Vocalizations." *The Journal of Experimental Biology*, October 21, 1996, 421-31, 421.

[230] *Noise and Health - Effects of Low Frequency Noise and Vibrations: Environmental and Occupational Perspectives* (PDF Download Available). Accessed Feb 17 2018 https://www.researchgate.net/publication/258400137_Noise_and_Health_-_Effects_of_Low_Frequency_Noise_and_Vibrations_Environmental_and_Occupational_Perspectives.

[231] Lee, Matthew, and Michael Weissenstein. "Hearing Loss of US Diplomats in Cuba Blamed on Covert Device." AP News. August 10, 2017. Accessed May 14, 2018. https://www.apnews.com/51828908c6c84d78a29e833d0aae10aa.

[232] Lederman, Josh. "19 American Diplomats in Cuba Suffering Health Problems after 'attacks' Blamed on Secret Sonic Weapon." The Independent. September 02, 2017. Accessed May 14, 2018. https://www.independent.co.uk/news/world/americas/us-diplomats-cuba-havana-embassy-deaf-secret-sonic-device-attacks-headaches-concussion-brain-injury-a7925376.html.

[233] Gao, Charlotte. "Is China Really Targeting US Diplomats With a 'Sonic Attack'?" The Diplomat. May 29, 2018. Accessed June 24, 2018. https://thediplomat.com/2018/05/is-china-really-targeting-us-diplomats-with-a-sonic-attack/.
[234] "Cognitive Disruptions." The Dietary History. Accessed December 30, 2018.
http://www.sharinginhealth.ca/presentations/cognitive_disruptions.html.
[235] Prihoda, Pearl J. *Man-Otang: Bigfoot Myth or Reality*. Bellevue: SAFEHouse Publishing, WA, 2014, 212.
[236] Ibid., 212.
[237] https://www.youtube.com/watch?v=DY1pcEtHI_w.
[238] https://www.alashensemble.com/about_tts.htm.
[239] https://video.nationalgeographic.com/video/exploreorg/inuit-throat-singing-eorg.
[240] https://www.youtube.com/watch?v=qnGM0BlA95I.
[241] "Piezoelectricity." Wikipedia. June 26, 2018. Accessed July 01, 2018. https://en.wikipedia.org/wiki/Piezoelectricity.
[242] Whale, Sharon. "The Piezo and Dielectric Properties of Crystals & Gem Stones." Whale Medical. Accessed July 01, 2018.
http://whalemedical.com/the-piezo-dielectric-properties-of-crystals-gems/.
[243] Royal Society of Chemistry, Staff. "Potassium - Element Information, Properties and Uses | Periodic Table." *Royal Society of Chemistry - Advancing Excellence in the Chemical Sciences.* Accessed July 14, 2018. www.rsc.org/periodic-table/element/19/potassium. The Royal Society of Chemistry members "are the world's leading chemistry community, advancing excellence in the chemical sciences."
[244] Royal Society of Chemistry, Staff. "Rubidium - Element Information, Properties and Uses | Periodic Table." *Royal Society of Chemistry - Advancing Excellence in the Chemical Sciences.* Accessed July 14, 2018. www.rsc.org/periodic-table/element/37/rubidium. The Royal Society of Chemistry members "are the world's leading chemistry community, advancing excellence in the chemical sciences."
[245] Staff. "Electron" Definition and Meaning | Collins English Dictionary. *Complacent Definition and Meaning | Collins English Dictionary.* Accessed July 15, 2018.
www.collinsdictionary.com/dictionary/english/electron.

[246] "Magnetic Moment." *Wikipedia*, Wikimedia Foundation, Accessed July 15, 2018. 13 July 2018, en.wikipedia.org/wiki/Magnetic_moment.
[247] "Paramagnetism." Wikipedia. July 23, 2018. Accessed July 27, 2018. https://en.wikipedia.org/wiki/Paramagnetism.
[248] "Electron Spin Resonance: Definition and Meaning | Collins English Dictionary. *Complacent Definition and Meaning | Collins English Dictionary*, www.collinsdictionary.com/dictionary/english/electron-spin-resonance. Accessed July 15, 2018.
[249] Prihoda, Pearl J. *Man-Otang: Bigfoot Myth or Reality*. Bellevue: SAFEHouse Publishing, WA, 2014, 188.
[250] Ibid., 188-189.
[251] NASA, Staff. "Activity II: Modeling Radiation-Damaged DNA." *Pdf/551392main_SF_Modeling_Radiation*, NASA - National Academy Press, Washington, DC., 1996, ISBN 0-309-05326-9, 1996, www.nasa.gov/pdf/551392main_SF_Modeling_Radiation.pdf. Accessed July 16, 2018.
[252] Brown, Kristen V. "Forget the Double Helix-Scientists Discovered a New DNA Structure Inside Human Cells." *Gizmodo*, Gizmodo.com, 23 Apr. 2018, gizmodo.com/forget-the-double-helix-scientists-discovered-a-new-dna-1825476635. Accessed July 16, 2018.
[253] "NASA could ALTER astronauts' DNA before sending them on Mars mission." Express. October 9, 2017. Accessed July 27, 2018. https://www.express.co.uk/news/science/864194/Mars-planet-NASA-insight-astronaut-DNA-cosmis-radiation-space-mission-red-planet.
[254] Prihoda, Pearl J. *Man-Otang: Bigfoot Myth or Reality*. Bellevue: SAFEHouse Publishing, WA, 2014, 187.
[255] Ibid., 189.
[256] Queensland Brain Institute, Staff. "What Are Neurotransmitters?" *Queensland Brain Institute*, The University of Queensland, Australia, 9 Nov. 2017. Accessed July 14, 2018. qbi.uq.edu.au/brain/brain-physiology/what-are-neurotransmitters.
[257] Prihoda, Pearl J. *Man-Otang: Bigfoot Myth or Reality*. Bellevue: SAFEHouse Publishing, WA, 2014, 166.
[258] Ibid., 166.
[259] Ibid., 182.

Chapter 17

Suppositions

"It's more fun to arrive at a conclusion [supposition] than to justify it."

— Malcom Forbes

This research did not begin with a hypothesis. Rather, it started with a highly unusual anomalous experience in 1999, which ultimately triggered nearly two decades of literature review and field research efforts. Although I stumbled and bumbled my way through in the beginning, I finally organized my thoughts, intent, and needs such that I was able to successfully pursue ongoing, valid sasquatch research.

It is difficult to present rock-solid conclusions about sasquatches because so much is still unknown. That is why I named this chapter Suppositions; and, of course, a fine synonym for the word supposition is *guesses*! Almost too droll for words, albeit far from being a travesty, eh?

Over the past 17+ years of research, I developed proposals about sasquatches. 25 of these proposals were presented in this book. They appear here in the order in which they were presented within the book. You should readily glean my train of thought about sasquatches and sasquatch behaviors by reviewing them. Following the list of proposals, I offer concluding thoughts on anomalous experiences, cognitive dissonance, coherence, a bit on how we choose an area for research, and, in closing, share my views on the question of species with respect to sasquatches.

I shall endeavor to maintain an open mind as the research continues—and continue it shall! In the interim, I trust that the material presented in this book will add a bit of insight into the overall shadowy unknowns about our inscrutable sasquatches.

25 Proposals About Sasquatches

Proposal 1:

> Changing Earth magnetism affects humans and all living organisms—including sasquatches.

Proposal 2:

> Solar and geomagnetic storms affect humans and all living organisms—including sasquatches.

Proposal 3:

> Changing Earth magnetism has a noticeable effect upon the energy emanating outward from the physical heart—both for humans and sasquatches.

Proposal 4:

> The sasquatches with whom I interact show an ability to think and employ at least an early form of reasoning.

Proposal 5:

Sasquatches use rocks, stones, sticks, etc., as forms of communication.

Proposal 6:

Sasquatches have strong observational skills, an aptitude to assess time in some manner, the skill to mimic human behavior, and can use discernment.

Proposal 7:

It is possible for a conjoined state of coherence to occur between a human and a sasquatch.

Proposal 8:

Geomagnetic storms offer sasquatch researchers' opportunities to engage in deliberate use of changing Earth and geomagnetic energies while observing and intentionally interacting with sasquatches.

Proposal 9:

When I engage in intentional coherence, I choose precisely how and where the electromagnetic energy waves emanating from my heart direct outward toward sasquatches.

Proposal 10:

The geology of the Laurentian Shield offers sasquatches a natural geomagnetic/electromagnetic habitat in which to wander and engage in typical sasquatch behaviors.

Proposal 11:

The geology of the Laurentian Shield offers a way for sasquatches to interact with and manipulate extremely low frequencies to their benefit.

Proposal 12:

The way magnetic senses have been proven to aid mammals and insects may be applied fairly and confidently to sasquatches.

Proposal 13:

Magnetite nanoparticles within the human brain may be manipulated by sasquatches controlling and directing extremely low frequencies.

Proposal 14:

As carbon-based creatures, sasquatches emanate a heart magnetic field—as do humans—and, quite possibly, their heart's magnetic field extends farther out than the 10 feet measured on humans.

Proposal 15:

Sasquatch researchers who learn the act of coherence can use directed intent to control their interaction with sasquatches.

Proposal 16:

Sasquatches—hairy, human-like apes—must be included in any discussion about biological systems involving electromagnetic energies.

Proposal 17:

It's possible for sasquatches to manipulate electromagnetic energies to intentionally affect human autonomic nervous systems, brains, and cardiovascular systems for purposes of self-defense or other reasons.

Proposal 18:

Sasquatches can manipulate plasma, thereby causing distortions in photos and videos.

Proposal 19:

Sasquatch eye glowing incidents link with energetic waveforms connecting infrared frequencies and magnetoreception qualities as they interact with Earth's changing magnetic field.

Proposal 20:

Electromagnetic induced transparency, waves and quantum interference, infrared frequencies, infrasound, piezoelectricity, electron spin resonance, radiation, and PPIT work together to allow sasquatches engagement in natural behaviors such as full induced or partial induced transparency.

Proposal 21:

Sasquatches use infrasound as part of their communication repertoire, most likely for defense or to stay in touch with other members of a family, tribe or community .

Proposal 22:

Sasquatches emit low frequency sounds around humans to intentionally create a sense of cognitive disruption or other types of dysfunction in humans, thereby effectively chasing them away from their territory.

Proposal 23:

Sasquatches use tapping, wood banging, rock banging, or other methods to create sound waves to alter natural electromagnetic qualities of quartz crystals, creating ELF electromagnetic energies to use for their specific benefit, quite possibly to assist with induced transparency.

Proposal 24:

When hydrogen sulfide (H2S) is intentionally manipulated by sasquatches—as suggested by Prihoda—they can emanate an unseen odoriferous miasma to disorient humans or animals, which serves as a nasty yet effective defense mechanism.

Proposal 25:

Sasquatches use a combination of electromagnetic light waves, chemical interactions, and electron spin physics to move into and/or out of a specific location unseen; this is another form of induced transparency.

Wildlife Use of Magnetism and Geomagnetism

Well known evidence exists to prove that wildlife engages in daily interactions with both magnetism and geomagnetism. From a scientific perspective, this is irrefutable; from a personal perspective, as always, you *choose* what you wish to believe irrespective of the evidence.

A tried and sure avenue leading to the realm of sasquatches is an open mind. Unfortunately, due to the ongoing hype and marketing of the many and varied televised sasquatch hunters, combined with silly, baroque commercials about sasquatches, being open minded about them can be a bit challenging, although definitely a requirement if one wants to test the waters of truth presented by dedicated sasquatch researchers.

Science has proven that flora and fauna interact with magnetism and geomagnetism. The data I've presented throughout this book confirms, at least in my opinion, that sasquatches also interact with magnetism and geomagnetism—oftentimes intentionally. Furthermore, I believe ongoing pursuit of the idea that sasquatches manipulate certain electromagnetic frequencies and light waveforms and, in some instances, particles of plasma, presents as a worthwhile ongoing research endeavor. In addition, I'm also convinced that anyone following the patterns of research outlined in this book could reproduce much of the evidence I've presented. To me, this is incredibly exciting!

I also believe that measuring magnetism and geomagnetism while documenting changes occurring over time—intermixed with intentional personal coherence—is a valid method to employ for successful sasquatch interactions. Of course, my experiences show habituation and familiarization likewise play significant roles in creating a space where intentional sasquatch interactions occur.

Induced Transparency Thrives in the Wilderness

In my worldview, induced transparency—by whatever means or whatever name—is positively, absolutely, smack-dab in your face real! So there! See what I mean about beliefs?

Goodness! Beliefs can be such albatrosses at times, eh? Or, perhaps a platform from which to dive knowingly into the enigmatic realm of sasquatches. For instance, there is nearly zero chance you could convince me that some form of naturally induced transparency does **not** occur. I and Brushtracker **saw** naturally induced transparency several times over the years. In addition, I've had healthy ongoing debates with myself about it. Hmm … that might be a little scary, eh?

Furthermore, I've discussed the issue with Maggie and Julie ad nauseum. In the end, I steadfastly believe that some form of induced transparency is real, and that it occurs as a *natural phenomenon* in contrast to a paranormal activity. Again, though, I do not dispute that paranormal activities occur. I merely claim that sasquatch transparencies I've seen occurred as a natural phenomenon.

I also categorically state that Pearl Prihoda's PPIT theory presented the sasquatch community with an electrifying realization that *induced transparency is real and active within the wilderness.* Although she and I moved toward a similar goal, we traveled somewhat dissimilar trails to arrive at similar conclusions.

My research offers evidence for a form of electromagnetic induced transparency. Moreover, it strongly corroborates Prihoda's meticulous and lengthy pursuit into this unusual and captivating aspect of sasquatches. Sadly, I confess to a deficit in my knowledge of chemistry. This simply means, though, that I need to restudy the area; that way, I can eventually comment semi-intelligently on Prihoda's use of chemistry in her hypothesis. I certainly intend to do this as my research moves forward!

Habituation, Familiarity, and Intentional Sasquatch Interactions

I began to recognize the ideas of habituation and familiarity from 2011 onward. My nearly six-week research journey in 2017 proved unquestionably that some sasquatches tend toward *intentional* non-hostile interactions.

Again, I do not bait them. I do not call them. I do not try to track them to whatever hidden places they may go. I do not hunt them. Instead, I strive simply to learn about them only as they are willing to interact with me. For example, I study what their movement patterns are; how they interact with their young; how they communicate; how they use magnetism and geomagnetism to their advantage; and so forth. My primary goal as a sasquatch researcher is to discover ways to positively interact with these inscrutable, intelligent, shy, and enormous beings of the wilderness such that some of their many mysterious qualities manifest as new or different knowledge to both science and the general population.

To a small degree, I feel I've been successful. To a larger degree, I have so much yet to learn, mistakes to make, hopes to dash, and forthcoming successes yet unrealized. I shall, however, continue taking magnetic measurements and employ tactics of habituation and familiarity with all future research endeavors, deeply intermingled, of course, with intentional coherence.

Deliberate Coherence and Intentional Sasquatch Interactions

Psychological effects stemming from an anomalous experience can lead to temporary cognitive dissonance. I have shown that adverse effects caused by cognitive dysfunction of any sort while in the field can be improved or avoided simply by reaching and sustaining *intentional, directed coherence.*

I have recurrently lived that axiom, not only in the field, but in many areas of my life. Seriously! Simply give it a try, remembering results may not be what you hope for at the outset. Why? Because becoming adept with successful, intentional coherence needs a good bit of practice, not to mention a possible massaging of your typical beliefs. I assure you, though, it is well worth whatever effort you put into reaching and sustaining intentional coherence. My results in the field after consistently employing this technique over the years has proven to be phenomenal.

Real rewards become self-evident when this technique is applied successfully in the field. Once again, you can readily prove this to yourself; simply learn the act of intentional coherence and then apply it while out in the wilderness seeking positive interactions with sasquatches. The results are sure to amaze and mystify you!

How we Choose an Area for Research

I engage in a specific routine when examining an area for research. Although the following list is numbered, I do not always adhere to the order shown, so please do not be misled by the numeric ordering. I do, however, *thoroughly* evaluate each feature or characteristic listed before moving into a new area.

1. Check magnetic anomaly maps of the area.
2. Check geographic and bedrock geology maps of the area.
3. Check hydrology maps of the area.
4. Check waterways in the area, including lakes, rivers, creeks, streams, falls, ponds, swamps, etc.

5. Check flood maps for the area.
6. Check flora and fauna in the area.
7. Check fault maps for the area.
8. Review topographic maps of the area to figure out elevation spread and diversity.
9. Review gravity maps of the area.
10. Research anthropological finds in the area.
11. Review population statistics of the area.
12. Determine agricultural statistics for the area.
13. Determine historic weather patterns for the area.
14. Determine historic space weather patterns for the area.
15. Ascertain hunting seasons and patterns for the area.
16. Determine off-road vehicle patterns for the area.
17. Identify federal, state, and local campsites within the area.
18. Find and record gravel pits in the area.
19. Find and record clean water sources in the area for our own consumption.
20. Determine the closest medical facilities within the area.
21. Determine the closest food access points in the area.
22. Research known historic sightings reported for the area.

The Question of Species

There is much discussion and debate among members of the sasquatch community about the question of sasquatch species. To begin, I do not personally hold to a single theory about the species issue. Why not? Because I simply do not know what they are or from whence these mysterious beings originated.

I do, however, believe that a lineage exists for sasquatches; one which likely follows the well-known Linnaean model of taxonomy for categorizing families, genus, and species. From what I've read—combined with firsthand experiences—I strongly suspect that there are many species of so-called sasquatches (regardless of name) around the world. Moreover, I believe distinct species not only look different from one another, each species undoubtedly acts differently from others. That is, I believe that each species exhibits innate personality traits both individually and collectively. I also suspect that each species engages in diverse forms of societal norms within their tribes, communities, territorial regions, or families.

Some species seem genetically gentle and curious, while others may be naturally surly and aggressive. Some species are physically small, while others are larger to much larger. Some species give the impression that they are more compatible with humans than others. Some are … but wait! I'm convinced that far too many variables are involved to attribute narrowly defined characteristics to an aggregate genus or species of sasquatches.

I'm pleased beyond belief, however, that the sasquatches with whom we interact appear to **not** be surly or aggressive! I do, though, believe they could be if brutally annoyed. I shall forever strive to avoid that unpleasant situation!

I suspect Loren Coleman and Patrick Huyghe are assuredly on the right track with the information they published in their 2006 edition of *The Field Guide to Bigfoot and Other Mystery Primates*. They suggest, for example, "a classification of eight possible mystery primates."[260] While I can neither confirm nor deny this, I certainly would not be surprised if their assessment turns out to be correct—or at least within an estimable ballpark.

Some believe the sasquatch is a knuckle-dragger descended from the *Gigantopithecus*, "an extinct genus of ape that existed from perhaps nine million years to as recently as one hundred thousand years ago, in what is now India, Vietnam, China and Indonesia, placing *Gigantopithecus* in the same time frame and geographical location as several hominin species."[261] Others suggest a direct relationship with Orangutans. Countless other ideas on the genealogy of sasquatches also permeate the sasquatch community.

Again, I have no clue what sasquatches are or what their origins might be. I do, however, believe unequivocally that the sasquatches with whom we interact are much closer to human than they are to animal. I believe this because I've seen a deep level of intelligence in their eyes and have had rather touching experiences with them. For example, recall the grapes and pin cherry pits incident described in Chapter 7.

I've also seen sasquatches show an affinity for reason. I've seen them act toward each other in a caring, loving manner. I've seen them show a curious interest in communicating with humans. I've had them share gifts with me. Because of these and many other reasons, I hold a high confidence that the sasquatches within our purview of familiarity **must** be closer to human than animal.

Hypothesis for 2018 Onward?

I've created a hypothesis for 2018 onward. Hold on to your hat, though, because it's a bold one. The null hypothesis is obvious from the hypothesis.

Hypothesis

> Habituation, familiarity, and coherence will prove proposals 1-25 during future field research.

Null Hypothesis

> Habituation, familiarity, and coherence cannot prove proposals 1-25 during future field research.

Final Thoughts

I'm extraordinarily confident that one day soon the sasquatch community will have answers to the many questions about the sasquatches discussed in this book. Not only that, I strongly believe that answers to questions not yet asked are also just around the corner of sasquatch reality.

I only hope that the answers arrive through development of a relationship with them rather than through aggressive means. "Dame

Jane Morris Goodall, formerly Baroness Jane van Lawick-Goodall, is a British primatologist and anthropologist."[262] She became world-renowned for her work with chimpanzees. If we take nothing else from her life's work, let us learn that we have hope for discovering ways to interact with sasquatches—entities less human than we—while using compassion, dignity, and integrity. Goodall led the way with this, and sasquatch researchers can certainly follow her poignant lead.

I fully intend to use habituation, familiarity, and coherence in my ongoing research efforts. I truly believe this approach opens a door to studying these mystifying, enigmatic creatures without harm to self or them.

Future research efforts will prove or disprove this belief; it's really that simple!

May the gentle nature in you continue to seek the gentle nature of our shadowy, intelligent, and inscrutable sasquatch friends.

Chapter 17

[260] Coleman, Loren, and Patrick Huyghe. The Field Guide to Bigfoot and Other Mystery Primates. Anomalist Books, 2006, back cover.

[261] "Gigantopithecus." Wikipedia. July 08, 2018. Accessed July 23, 2018. https://en.wikipedia.org/wiki/Gigantopithecus.

[262] "Jane Goodall." Wikipedia. Wikimedia Foundation, March 15, 2019. https://en.wikipedia.org/wiki/Jane_Goodall.

Chapter 18

2018 and Beyond

"Action precedes change—act wisely and change intentionally!"

— Roger

This saga of *Sasquatch and Me* is far from over. In fact, I dare say a new and exciting beginning is unquestionably on the threshold, jauntily peeking at me from just over the wavering mists of the horizon. I also believe that the eyes of success seem to be shining encouragement on the many dedicated, diligent, persistent sasquatch researchers from around the world.

I'm confident many others feel something similar, for I am not the only one holding a strong belief that quests to uncover the mysteries surrounding bipedal anomalies is nearing a stage of fruition—or at the least, will soon take a huge step forward in peeling yet another layer of the multi-layered mysteries surrounding so-called sasquatches.

I truly believe the sasquatch community is on the cusp of extraordinary understandings about this relatively unknown species that has effectively evaded scientific verification for a century or more. With a bit of persistence, continued diligent literature research, a willingness to try innovative field research approaches, and, of course, a gentle touch by that *faerie* known as fate, I suspect a bipedal anomaly recognized as sasquatch will, sooner rather than later, be introduced into our societal awareness as a scientific reality.

Of course, we would then need to address a trainload of issues that, until now, have been considered as purely hypothetical. For example, how will the scientific and political communities deal with that eventuality? Hmm! I singularly shiver with excitement and shudder with concern while striving to perceive that prospect. Truly, there is much to ponder as we move ever closer to confirming the reality of the mysterious and elusive sasquatches.

Why We Continue Our Quest into the Unknown

Hmm! This is both **easy** and **difficult** to describe.

EASY, because from the first introductory moment in 1999, my life changed radically and instantaneously. How could it have done otherwise? Those who have had similar experiences know very well the answer to this question. On the other hand, those who have not shared such experiences will surely find it difficult to grasp the depth of meaning or the profound effect such experiences have on one's psyche.

From the original unexpected and insidiously odoriferous introduction to the first sighting of a large hairy, human-like ape an hour or so later, a desire to quantify the experience as real and not nuts exploded from deep within me. Once that turmoil settled down a bit, I began to nurture an unfathomable and indefinable motivation to seek answers which seemed not to exist.

For almost two decades, a nearly insatiable desire to learn to interact with this unclassified species took front and center in my awareness. Literally, each year's research experiences and results, combined with many challenges and ofttimes hard-learned lessons, added fuel to that deep desire and motivation initially triggered by an unsought, unforeseen experience.

From then until now, the ride has been amazing, sometimes tumultuous, other times calm to the point of serene, always educational, and complete with a rainbow of successes and failures. Of course, I really do not believe in *failures*; rather, I view each "miss" as a lesson learned or merely a mistake not to be repeated.

DIFFICULT, because we are complex, intricately designed, carbon-based beings of consciousness made up of physical, emotional, mental, and many claim, so-called spiritual attributes. As such, until we fully understand *all* aspects of what we are, how we function, why we respond/react the way we do in irregular situations, conditions, and environments, and what our overall relationship is within the vast reaches of what we call universes, we are not likely to uncover ALL the definitive empirical evidence that we, as a human race, strive so diligently to expose.

From September 1999 through the present, though, I have stumbled and bumbled my way into a realm of reality that—prior to 1999—was a mere arbitrary conceptual possibility. Since that first physical experience, I have both intentionally and inadvertently had a variety of interactions with *something* which continues to defy precise explanation.

I have read thousands of pages of information written by folks doing their best to research and describe this inexplicable phenomenon according to their interpretations. I have also read well over a thousand witness reports and viewed hundreds of photos and videos. I have gathered research data leading to both clear and obtuse speculation about some of the mysterious qualities attributed to the seemingly unfathomable being often called sasquatch. I have pondered all I've seen and have had occasion to briefly study a few of these shadowy entities visually or have otherwise interacted with them through supplementary means. Even so, I remain knowing very little about these inscrutable beings known as sasquatches.

I've done all those things and more for nearly two decades. Yet, I cannot offer absolute empirical proof for their existence. Even so, I continue to hold strongly to my belief that these secretive creatures are real and one day soon their realness shall enter the empirical realm of scientific authentication.

I have not in the past, do not now, nor shall I in the future, purport that I will be the one to deliver such scientific proof. Rather, I'm persuaded by all I've read, seen, heard, and experienced that someone somewhere, sometime will soon "flip a switch" and shine a bright light on our sasquatch ignorance. Some event will occur that will convince the world not only of their existence, but of the necessity to learn all we can—not only about them; rather, from them. Of course, at that time, society will struggle with myriad ethical issues that eventual recognition of this species produces, including how to preserve wilderness areas as habitats for their long-term well-being.

In the end, this story of sasquatch and me is really nothing more than a tiny seed planted where earlier untold seeds had taken root within our awareness; seeds planted by many courageous trailblazers who came long before silly folks like me stumbled unwittingly onto the trail of so-called sasquatches.

Minus the gutsy pioneers who bravely confronted stark criticisms or ongoing storms of public scorn, none who write today about the mysterious sasquatches could do so fruitfully without the foreknowledge gleaned from our stalwart predecessors. For me to not remember these pioneers who valiantly blazed a way into the unknown realm of sasquatches would be an undying shame.

The early researchers, many who devoted a *lifetime* of personal sacrifice to their investigations and—regardless of any mistakes made—blazed a path for those of us who follow in their gigantic footsteps. We must stay grateful for their courage and personal sacrifices. We must acknowledge and honor their efforts, both individually and collectively. We must not forget that without them and their work we would indeed have a precarious foundation upon which to build current theories or share delightful and meaningful stories and experiences.

We who follow must realize each of us merely adds to an ever-growing database of information about our puzzling sasquatch friends. I believe when all is said and done, it is through the sasquatch community's **cumulative** effort that we will eventually pull the final veil away from the mystery of our unfathomable sasquatches.

That said, my ultimate hope is that this story of sasquatch and me might, in some small way, add to the whole of evidence that already exists. Why? Because I'm a strong proponent of the idea that each who plants another seed in the beautiful and growing meadow of accumulated knowledge about sasquatch research makes a significant contribution to an eventual finished product: *Sasquatch Unveiled.*

Lloyd Pye Fund for Alternative Research

I received a grant from the **Lloyd Pye Alternative Research Foundation** to assist with my 2018 research. I am indebted to Amy, Nina, and The Lloyd Pye Fund for Alternative Research board for this generous grant as well as their belief in my work—regardless of how elusive results from this type of research can be at times.

A point of note: As part of the Lloyd Pye grant agreement, I will be writing a chapter for a book that the Lloyd Pye Fund for Alternative Researchers intends to publish about the various types of alternative research they help fund.

I dearly hope my 2018 research efforts do not disappoint them. Thank you so much! Also, I wish I could have shared this story of Sasquatch and Me with Lloyd. I suspect he would have enjoyed it. Amy and Nina, I will not let you down.

I very much look forward to my next research trip … and the next … and the next … and … with a deep appreciation for the experiences I've had to date interwoven with a never-ending anticipation that marvelous, revealing adventures are imminent, lying teasingly just through the ever-sparkling mists along the horizon. I don't mind being teased a bit as I continue to pursue my quest, because the quest itself is a large part my personal reward.

I honor all who have in the past, are now, and will in the future plant their personal seed(s) of knowledge into the ever-growing database of information about this seeming inexplicable mystery. I offer a humble and grateful shout-out to all who have, continue to do so now, and will in the future aid in expanding the beauty lying in that meadow of sasquatch knowledge, and hope each day to watch this knowledge bloom into scientific validation of these wondrous forest creatures that captivate the imaginations of so many.

This project—from 17+ years of research to this book—has truly been one of the greater undertakings in which I've engaged this lifetime. Nonetheless, I've enjoyed it immensely and shall continue to do so far into the future!

So many choice locations to return to; so many sites not yet explored. Yes, indeed. This saga of *Sasquatch and Me* is undeniably far from over! Until the future arrives and produces amazing results worth noting …

Go you intrepid Sasquatches!

Chapter 19

What We Believe About Sasquatch

"Knowledge of other people's beliefs and ways of thinking must be used to build bridges, not to create conflicts."

— Kjell Magne Bondevik, 26th Prime Minister of Norway

This chapter presents some of my beliefs regarding sasquatches, addressing the logic and cogent thought underlying them when possible. While, for the most part, what I believe about sasquatches is pure conjecture, I propose that the many encounters I've had with this species during almost two decades of dedicated private research qualifies me to speculate with at least a modicum of credibility.

This chapter is laid out with headings that name a common descriptive approach to sasquatch beliefs in general. Under each heading I state what I believe about that topic, offering evidence when available.

Community

I start with community because of the many photos I've taken of what I refer to as community bedding areas. I have presented several of these photos throughout the book. What I call community, many others in the field of sasquatch phenomenon call tribes. I believe either works to portray the intent; that small groups of sasquatches will gather to form a community or tribe. One likely reason for this is safety in numbers; another probable reason is companionship. After all, humans gather for similar reasons. I'm sure there are many other possible reasons as well; however, these are the two that strike me as relevant here.

I also believe that within a community, families hang together as subunits, gathering as a community for, perhaps, distinct social reasons. When the reason for gathering as a larger community concludes, I suspect that families within the community separate, although they likely stay within physical regional boundaries set up by the community in general. I might label these boundaries "community boundaries," which probably exist within larger regional boundaries, and so forth.

Communication

I believe that communication occurs between sasquatches on a regular basis. I do not know if the genera itself has a language, or if species within it have distinct languages. I have, however, seen many instances where credible evidence points strongly to those sasquatches that I've interacted with having a clear ability to communicate amongst themselves. Moreover, a bit of evidence for sasquatches having a language does appear within the sasquatch arena. For example, R. Scott Nelson, who has a background in deciphering cryptic coding associated with communication in Naval exercises, tackled the task to decipher the well-known Berry/Morehead sound tapes (tapes containing alleged sasquatch communication). Nelson produced his results in a study that leans heavily on phonetics as a basis for language.[263]

Maggie , Julie, and I have personally heard a variety of sounds emitted by sasquatches. Of course, Brush-tracker, Dragonfly, and BoB have as well. We have heard articulated whistles of variant pitches, tones, length, and visceral qualities associated with some of the vocalizations. We have also heard chirping sounds, guttural sounds, grunting sounds, and chortles or chuckles. I have heard what I believe is a low-level laugh, or certainly something similar. I likewise once heard an incredibly intimidating, frightening scream—**a sound one never forgets**.

We have heard something known as wood knocking many times. Wood knocking is a phenomenon many within the sasquatch community associate with a large stick banging against a tree, two sticks banging against each other, or rocks banging out single, or numerous sounds. Different patterns of sound associate with wood knocking and, it seems to me that different patterns of sound reflect specific types of communications. For example, one pattern might mean danger, while another might mean check out these crazy damn humans! On the other hand, as described in Chapter 13, the wood knocking symphony Julie and I heard near the end of our 2017 research trip sounded almost magical, certainly musical, and resounded through the forest as if the sounds had joined in concert with the subtle, unseen movement of all wilderness sound waves.

Oddly enough, at the end of the 2017 trip, I even saw sasquatches attempt communication with me using unusual hand and arm signals. This must be another form of communication. In turn, I too used basic, or, more likely, crude hand signals in response to them. No one was hurt because of this arm, hand, finger signaling so maybe they understood something I tried to portray even though I remain in the dark as to what they wanted me to understand. **Bottom line:** I believe there is undisputed evidence pointing to sasquatches engaging in variant forms of communication. I, however, am not able to verify that the communication I've experienced is a language of sorts.

Diet

The sasquatches found in the areas we research are likely omnivores, with a strong tendency toward being predominantly herbivores. They certainly eat a lot of wild berries and fruit, and we have seen no evidence of meat eating other than grubs or, perhaps small rodents. Of course, this in no way suggests that the sasquatches with which we interact do not eat meat.

I can say, though, that blueberries, huckleberries, raspberries, and blackberries are prevalent in our area of research. The area also supports wild mulberry, wild cherry, wild apple, wild plum, and wild downy juneberry trees. In addition, hundreds of species of mosses, lichens, liverworts, ferns, grasses, sedges, orchids, wildflowers and shrubs grow abundantly in our research area.

A note on lichens: Lichens are "a composite organism made up of a fungus in a co-operative partnership with an alga. Owing to this partnership, lichens can thrive in harsh environments such as mountaintops and polar regions. Characteristically, [lichens] form a crust like or branching growth on rocks or tree trunks."[264] As an interesting aside, "*Bryoria* is a common genus of lichen across the United States. In times of hardship, some First Nation tribes would eat this lichen, while other tribes sought it out. Some lichens were fed to pets during hard times as well."[265] Finally, "It is worth noting that a given forage may appear to be a rather small percentage of an animal's yearly diet yet play an important strategic role [in their diet]."[266]

There are at least eight varieties of wild mushrooms that grow in our research area: morel, oyster, honey fungus, combs tooth, sulfur shelf, hen of the woods, and chanterelle. There is also wild asparagus and watercress, not to mention plentiful wild nuts, such as walnuts, hickory nuts, acorns and other similar nuts from trees and shrubs.

You can also find wild onions, wild leeks, and nodding wild onion plants throughout the Northern Highland geographical wilderness area. Common milkweed, toothwort, white trout lily, Jerusalem artichoke, and other wild flowers also appear in the area. There are likewise a few edible weeds. See "Native Plants for Edible Landscaping," available online at http://www.goodoak.com/info/EdibleWildPlants.pdf.

We have evidence that sasquatches may eat certain types of dirt for its nutritional value. For instance, one area we researched showed evidence of dirt being eaten (see Chapters 6-7). You may recall that a brother-in-law aided me in having that dirt tested at a Michigan State University Soil and Plant Nutrient Laboratory. The results showed that potassium occurred in the dirt, as well as magnesium and calcium. We have elsewhere sufficiently addressed the importance of these minerals.

Once again, although we have seen no evidence of such in our plentiful research areas, many reports suggest that sasquatches might also eat deer, fish, small rodents, etc. For example, the northern white cedar tree serves as a host plant for a variety of insects, and we have seen evidence of the sasquatch eating insects, grubs, etc. I wish to both reiterate and stress that simply because we have not seen sasquatches partaking of meats does not suggest they avoid it.

Odor

Many people lay claim to the idea that sasquatches sometimes emit an obnoxious, strongly offensive odor, or said differently, a putrefied nasty stench. A common theme within the sasquatch community says that this stench is intended to warn someone or something away from the sasquatch emitting the odor. You may recall from Chapter 1 that our very first encounter with a sasquatch came about after smelling a nasty stench. The stench is real, and my personal preference is to **not** smell it again! As a reminder, sasquatch researcher Pearl Prihoda believes that this stench relates to the way sasquatches interact with hydrogen sulfide.

Markers

I believe sasquatches leave a variety of directional markers throughout the forest. I also believe that these markers fall under the overall umbrella of communication. I have accumulated myriad photographs of stones, sticks, and trees we believe were used to mark specific areas or pathways. To wit, recall from Chapter 9 that stones were moved around Pyramid Rock to represent a "cat's paw." Sasquatches also use small sapling trees by bending them over to show a seeming important and/or specific mark for directional or explicit site marking purposes. These bent saplings are often woven together at their tips. This leads nicely to the next section, nesting.

Nesting

Nesting sites are a well-discussed topic in sasquatch lore. We have seen many areas over the years that assuredly, at least to me, qualify as so-called nesting sites. I have many photos substantiating this claim, although the nesting sites we have seen all appear to be temporary. Furthermore, while many within the sasquatch community suggest that sasquatches also nest in trees, we have not been privy to a nesting site in a tree, only those we've found on the ground.

Sasquatch researcher Pearl Prihoda wrote about nesting, saying, "All great apes make complex sleeping structures… All nests are temporarily occupied, some for afternoon naps while others may be used to spend the night in."[267] She also wrote, "Nest building is a taught technique and may take [juveniles] some time to master."[268]

The nesting sites we have seen range from quickly put together nests to far more elaborate ones. The most elaborate area we've seen and photographed included an area on the shore of a river, with a deep pine and deciduous forest next to it. This area had 16 bedding sites where the creatures seemed to lie face down with their knees bent to their chests and foreheads touching the ground while sleeping. The 16 bedding sites we saw and photographed were clearly indicative of this type of sleeping method. The entire horseshoe-shaped bedding area was surrounded with broken branches and twigs, suggesting that a warning system was in place if anyone or anything was to approach it. As a reminder, this site was discussed in Chapter 5.

Finally, we have seen several nesting sites where small saplings and aspens were bent down and woven together at the tops. We saw branches placed over these as well as woven between the saplings or aspens. These sites seemed to be made for a longer duration stay than an afternoon nap; they were probably used for overnight—or longer—stays.

Hostility

Throughout years of private, low-key research, we have not experienced hostility of any sort directed at us. We do not, of course, know what the future holds; however, we are hopeful that hostility does not visit itself upon us!

The closest to a hostile sound I've heard to date occurred in the Fall of 2003; I drove alone to an Upper Peninsula site in Michigan where I intended to sit and watch after dark. The time was shortly after 1:00 a.m. Central Daylight Time (CDT). As I stepped out of my vehicle, I heard an ear-splitting, nearly indescribable, and spine-tingling harrowing howl-screech. I perceived the sound to be from one-half to three-quarters of a mile from my location, although the reverberation through the forest from the sound waves was astounding. I at once stepped back into my vehicle; I heard the sound once more and decided to leave the area and return to watch another night!

I most assuredly would **not** want to be in the near vicinity of a creature making that sound! I suspect that scary sound is indeed intended to strike a sense of fear in the listener; therefore, I believe this sound stands for a hostile warning, or, perhaps, a rather stern notice that precedes a potential hostile act. Again, although it was distant enough from me that I did not feel it was directed at me and, I did not feel threatened, I bugged out anyway. I think this has something to do with the recognition that whatever made this inhuman ear-piercing howl-screech caused feelings of energetic reverberations to resound throughout my being. For sure and certain, it was not an entity offering to engage in a new friendship! In fact, I'm quite certain it was not going to request a cup of coffee!

Another time, while hiking an area that Brushtracker and I had been in a day before, we found an approximately 70-yard long stretch of destruction through a group of small aspen trees and large shrubs. It appeared as if something tore through the area while in a rage; I can assure you that whatever it was, it was exceptionally large and powerful. The destruction was clearly caused by a **large** physical being as it ripped through the area while swinging its arms to either side. This incident was discussed in Chapter 3. Again, I would not have wanted to be around while that type of rage was being expressed.

Migration

Many renowned researchers within the sasquatch community believe that sasquatches migrate. Although some migration theories I've read lend credence to this idea, I am unable to offer supportive field research evidence for a clear theory about migration. Even so, I hold strong suspicions that migration does occur, at least on a regional basis. At this point, I tend to adhere to longtime bigfoot researcher Bobbie Short's suggestion. She wrote a piece for her website on the theory of migration, "… I find myself leaning heavily toward a behavior that I call transient wanderings more than I do migration…"[269] Bobbie Short has, sadly, passed, leaving behind a legacy of thought-provoking, credible information on the sasquatch phenomenon.

A possible, and certainly interesting aspect of migration inserted itself into our Fall 2017 research and was discussed in Chapter 12. This aspect relates specifically to the idea of regional migration, or perhaps, regional transient wanderings when longterm water supplies change.

Ability to Reason

I discussed the idea of sasquatches and reason at some length in Chapter 5. Please refer to Chapter 5 for information about why we believe sasquatches show an amazing ability to reason.

Hibernation

I am not able to speak knowledgably to this topic. I have no personal experience with it because I have not been to my research areas during winter. Winters in the western part of the Upper Peninsula of Michigan and northern Wisconsin can be quite daunting. As such, a trip north during the winter necessitates equipment and transportation modes we do not yet have. It is certainly possible for sasquatches to hibernate for short periods. I find it challenging, however, to believe that they hibernate in a manner identical to bears.

Do Sasquatches Have So-called Dimensional Qualities?

Addressing this idea could easily lead one to purport that sasquatches have paranormal skillsets. I am unable to either support or refute this idea. Regardless of whether sasquatches engage in dimensional activities, I strongly believe that many so-called paranormal happenings are readily explained with a bit of **natural** quantum phenomena.

Now then, to be even more clear on this issue, I happen to believe that certain paranormal experiences are, in fact, possible. Hence, outright disbelief is not the issue for me. Rather, in this instance, I'm convinced that sasquatches have an ability to control certain quantum parameters that allow them to engage in behaviors that *appear* to be paranormal.

Chapter 19

[263] Nelson, R. Scott. "Bigfoot Language Study Is Released." Bigfoot Language. June 20, 2010. Accessed March 22, 2018. Available, http://www.nabigfootsearch.com/Bigfootlanguage.html.

[264] Arkive, "Wisconsin's Northwoods." Accessed 07 November 2017. http://www.arkive.org/eco-regions/wisconsins-northwoods/.

[265] United States Department of Agriculture (USDA), "Lichens—Did You Know?". Accessed 07 November 2017. https://www.fs.fed.us/wildflowers/beauty/lichens/didyouknow.shtml.

[266] Stephen Sharnoff and Roger Rosentreter (Bureau of Land Management, Idaho), "Lichen Use by Wildlife in North America." Accessed 07 November 2017. http://www.lichen.com/fauna.html.

[267] Ibid., 67.

[268] Ibid., 57.

[269] Bobbie Short, "Sasquatch Migration – No Shortage of Opinions." Accessed 06 December 2017. http://www.bigfootencounters.com/biology/migration.html/.

List of Photos

List of Diagrams

List of Tables

Selected Bibliography

Bayanov, Dmitri, Christopher L. Murphy, and Roger Knights. *Bigfoot Research: The Russian Vision*. Surrey, BC: Hancock House, 2011.

Bindernagel, John A. *The Discovery of the Sasquatch: Reconciling Culture, History and Science in the Discovery Process*. Beachcomber Books, 2010.

Bindernagel, John Albert. *North America's Great Ape: The Sasquatch: A Wildlife Biologist Looks at the Continent's Most Misunderstood Large Mammal.* Beachcomber Books, 1998.

Birn, Joachim, and Eric Priest. *Reconnection of Magnetic Fields: Magnetohydrodynamics and Collisionless Theory and Observations*. Cambridge: Cambridge University Press, 2007.

Blocksma, Mary. *Naming Nature: A Seasonal Guide for the Amateur Naturalist.* New York, NY: Penguin Books, 1992.

Bord, Janet, and Colin Bord. *Bigfoot Casebook Updated: Sightings and Encounters from 1818 to 2004*. Pine Winds Press, 2006.

Buhs, Joshua Blu. *Bigfoot - The Life and times of a Legend.* Chicago: Univ. Of Chicago Press, 2010.

Buydens, Sharon. *3 Bigfoot Sightings Maps in 7 States*. Sharon Buydens, 2017.

Byrne, Peter. *The Search for Bigfoot: Monster, Myth or Man?* Pocket Books New York, 1975.

Childre, Doc Lew, Howard Martin, Deborah Rozman, and Rollin McCraty. *Heart Intelligence*. Cardiff, CA: Waterfront Press, 2016.

Coleman, Loren, and Patrick Huyghe. *Field Guide to Bigfoot and Other Mystery Primates*. Anomalist Books, 2015.

Coleman, Loren. *Bigfoot: The True Story of Apes in America*. Paraview Pocket Books, 2003.

Coleman, Loren. *Tom Slick: True Life Encounters in Cryptozoology*. Linden Publishing, 2002.

Crowe, Ray. *Bigfoot Behavior: The Best of the "Track Record"*. Vol. 1. Rhettman A. Mullis, Jr., 2011.

Crowe, Ray. *Bigfoot Behavior: The Best of the "Track Record"*. Vol. 2. Rhettman A. Mullis, Jr., 2011.

Crowe, Ray. *Bigfoot Behavior: The Best of the "Track Record"*. Vol. 3. Rhettman A. Mullis, Jr., 2011.

Debenat, Jean-Paul, and P. H. LeBlond. *The Asian Wildman: Yeti, Yeren & Almasty: Cultural Aspects & Evidence of Reality*. Hancock House, 2014.

Drummond, Allan, and James Dodds. *Wild Man of Orford.* 3rd ed. Jardine Press, 2015.

Fagg, Lawrence W. *Electromagnetism and the Sacred: At the Frontier of Spirit and Matter.* Continuum, 1999.

Filler, Aaron G., and David R. Pilbeam. *The Upright Ape a New Origin of the Species*. New Page Books, 2007.

Forgey, William W., MD. *Wilderness Medicine: Beyond First Aid.* 6th ed. Falcon PR Pub, 2017.

Fusch, Ed. *Scweneyti and the Stick Indians of the Colvilles: The Interaction of Large Bipedal Hominids with American Indians.* 2nd ed. Riverside, WA: Ed Fusch, 2002.

Gloss, Molly. *Wild Life: By Molly Gloss.* Boston, MA: Houghton Mifflin, 2000.

Gooch, Stan. *The Neanderthal Legacy: Reawakening Our Genetic and Cultural Origins.* Inner Traditions, 2008.

Gormly, Mary, and Dmitri Bayonov. *Northwest Anthropological Research Notes.* 1st ed. Vol. 2. University of Idaho, Spring 1977.

Green, John. *On the Track of the Sasquatch.* Hancock House, 1994.

Green, John. *Sasquatch: The Apes among Us.* Hancock House, 1978.

Green, John. *The Best of Sasquatch Bigfoot: The Latest Scientific Developments plus All of On the Track of the Sasquatch and Encounters with Bigfoot.* Surrey, B.C.: Hancock House, 2004.

Green, John. *The Sasquatch File.* Agassiz, B.C.: Cheam Pub., 1973.

Green, John. *Year of the Sasquatch.* Agassiz, B.C.: Cheam Pub., 1970.

Guttilla, Peter. *The Bigfoot Files.* Santa Barbara: Timeless Voyager Press, 2003.

Heuvelmans, Bernard. *Neanderthal: The Strange Saga of the Minnesota Iceman.* Anomalist Books, 2016.

Heuvelmans, Bernard. *On the Track of Unknown Animals.* Routledge, 2016.

Hunter, Don, and René Dahinden. *Sasquatch: Bigfoot, the Abominable Snowman - An Ancient Myth or the Missing Link in Man's Evolution?* McClelland and Stewart, 1973.

Krantz, Grover S. *Big Foot-Prints: A Scientific Inquiry into the Reality of Sasquatch.* Johnson Books, 1992.

Krantz, Grover S. *Bigfoot Sasquatch Evidence.* Hancock House Publishers, 2008.

Lapseritis, Jack, Christopher L. Murphy, Lee Trippett, and Jesse D'Angelo. *The Sasquatch People and Their Interdimensional Connection.* United States: Comanche Spirit Pub., 2011.

Lapseritis, Jack. *The Psychic Sasquatch and Their UFO Connection.* CreateSpace, 2005.

Lester, Shane. *Clan of Cain: The Genesis of Bigfoot.* Shane Lester, 2001.

Lynch, Gary, and Richard Granger. *Big Brain: The Origins and Future of Human Intelligence.* New York, NY: Palgrave Macmillan, 2009.

Mack, Clayton. *Grizzlies & White Guys: The Stories of Clayton Mack.* Harbour Publishing, 1993. Compiled and edited by Harvey Thommasen; foreword by Mark Hume.

McCraty, Rollin. *Science of the Heart: Exploring the Role of Heart in Human Performance.* Vol. 2. HeartMath Institute, 2015.

Meldrum, Jeff. *Sasquatch: Legend Meets Science.* New York: Forge, 2007.

Merrill, Ronald T. *Our Magnetic Earth: The Science of Geomagnetism.* Chicago: University of Chicago Press, 2012.

Miller, Dorcas S., and Cherie Hunter Day. *Track Finder: A Guide to Mammal Tracks of Eastern North America.* Nature Study Guild, 1981.

Morgan, Robert W. *Bigfoot Observer's Field Manual: A Practical and Easy-to-Follow, Step-by-Step Guide to Your Very Own Face-to-Face Encounter with a Legend.* Pine Winds Press, 2008.

Morgan, Robert W. *Soul Snatchers: A Quest for True Human Beings.* Pine Winds Press, 2008.

Murphy, Christopher L., and Roger Knights. *Know the Sasquatch/Bigfoot: Sequel & Update to Meet the Sasquatch.* Surrey, B.C.: Hancock House, 2010.

Naegele, Thomas A., D.O. *Edible and Medicinal Plants of the Great Lakes Region.* Davisburg, MI: Wilderness Adventure Books, 1996.

Napier, John Russell. *Bigfoot the Yeti and Sasquatch in Myth and Reality.* E. P. Dutton and Company, 1973.

Napier, John Russell., and Russell H. Tuttle. *Hands.* Princeton University Press, 1993.

Paulides, David. *The Hoopa Project: Bigfoot Encounters in California.* Hancock House, 2008.

Paulides, David. *Tribal Bigfoot.* Crypto Editions, 2017.

Paäbo, Svante. *Neanderthal Man: In Search of Lost Genomes.* Basic Books, a Member of the Perseus Books Group, 2015.

Powell, Thom. *Edges of Science.* Willamette City Press, LLC, 2015.

Powell, Thom. *Shady Neighbors.* CreateSpace, 2011.

Powell, Thom. *The Locals: A Contemporary Investigation of the Bigfoot/Sasquatch Phenomenon*. Hancock House Publishers, 2003.

Pye, Lloyd. *Everything You Know Is Still Wrong*. Nina Pye and Amy Vickers, 2017.

Pyle, Robert Michael. *Where Bigfoot Walks: Crossing the Dark Divide*. Berkeley, CA: Counterpoint Press, 2017.

Quasar, Gian J. *Recasting Bigfoot: Uncovering the Truth About Sasquatch Amidst the Hype of Bigfoot*. Brodwyn-Moor & Doane, 2010.

Ratcliffe, J. A. *An Introduction to the Ionosphere and Magnetosphere*. London: Cambridge University Press, 1972.

Rezendes, Paul. *Tracking & the Art of Seeing: How to Read Animal Tracks & Sign*. New York: HarperCollins, 1999.

Sanderson, Ivan T. *Abominable Snowmen: Legend Come to Life*. New York: Cosimo Classics, 2008.

Shackley, Myra L. *Still Living? Yeti, Sasquatch, and the Neanderthal Enigma*. Thames and Hudson, 1986.

Sheppard-Wolford, Sali, and Autumn Williams. *Valley of the Skookum: Four Years of Encounters with Bigfoot*. Pine Winds Press, 2006.

Shiel, Lisa A., and Kerrie Shiel. *The Hunt for Bigfoot: A Novel*. Jacobsville Books, 2011.

Slate, B. Ann., and Alan Berry. *Bigfoot*. Toronto: Bantam, 1976.

Smith, Bruce D. *The Emergence of Agriculture*. Scientific American Library, 1998.

Strain, Kathy Moskowitz. *Giants, Cannibals & Monsters: Bigfoot in Native Culture*. Hancock House, 2008.

Suchy, Linda Coil, and Christopher L. Murphy. *Who's Watching You? An Exploration of the Bigfoot Phenomenon in the Pacific Northwest*. Surrey, B.C.: Hancock House, 2009.

Sykes, Bryan, and Tony Ricketts. *The Seven Daughters of Eve*. Royal New Zealand Foundation of the Blind, 2009.

Sykes, Bryan. *Adam's Curse: A Future Without Men*. Corgi, 2010.

Sykes, Bryan. *Bigfoot, Yeti, and the Last Neanderthal*. Disinformation Books, 2016.

Sykes, Bryan. *DNA USA: A Genetic Portrait of America*. Liveright Publishing, 2013.

Sykes, Bryan. *The Nature of the Beast: The First Scientific Evidence on the Survival of Apemen into Modern Times*. Coronet, 2015.

Thayer, Samuel. *The Forager's Harvest: A Guide to Identifying, Harvesting, and Preparing Edible Wild Plants*. Birchwood, WI: Forager's Harvest, 2006.

Weatherly, David. *Wood Knocks: Journal of Sasquatch Research*. Vol. 1. Leprechaun Press, 2016.
Weatherly, David. *Wood Knocks: Journal of Sasquatch Research*. Vol. 2. Leprechaun Press, 2017.
Williams, Autumn. *Enoch: A Bigfoot Story*. CreateSpace, 2010.

About the Author

Roger and Maggie live in Michigan; their six adult children and nine grandchildren are a major focus in their lives. Roger's varied background has prepared him for the type of research he so dearly loves. While this is his first book, he is not a stranger to writing. He has prepared and presented information on many topics over the years, including a variety of PowerPoint seminars and workshops across several states.

He is an avid admirer of the wilderness and spends as much time as he can enjoying all that the wilderness has to offer. As noted on the back cover of the book, he has been a journeyperson bricklayer, masonry contractor, restaurant owner, computer repair/maintenance business owner, instructor of digital electronics at a community college, electromagnetic research specialist, minister, spiritual counselor, business consultant, electronics enthusiast, and, of course, a passionate sasquatch researcher.

One of his favorite pastimes is researching. It was during one of myriad conversations with his dear friend Julie that a simple question from her inspired him to prepare and present this, his first book. This is not to suggest you should blame Julie if you absolutely hate it; rather, simply be aware that she was his original inspiration, with Maggie quickly following suit. Without their ongoing love and enduring encouragement, this book would not have been written.

You can reach Roger via email: 56647@pyfn.com, or through his website: sasquatchandme.com.

Index

1

2

A

B

C

D

G

H

I

N

O

P

Q

R

S

T

U

V

W

X

Y

G. Roger Stair

Made in the USA
Las Vegas, NV
03 November 2026

57422840R00274